MATLAB Deep Learning

Learning

With Machine Learning, Neural
Networks and Artificial Intelligence

Phil Kim

Apress®

MATLAB Deep Learning: With Machine Learning, Neural Networks and Artificial Intelligence

Phil Kim
Seoul, Soul-t'ukpyolsi, Korea (Republic of)

ISBN-13 (pbk): 978-1-4842-2844-9   ISBN-13 (electronic): 978-1-4842-2845-6
DOI 10.1007/978-1-4842-2845-6

Library of Congress Control Number: 2017944429

Cover image designed by Freepik

   Managing Director: Welmoed Spahr
   Editorial Director: Todd Green
   Acquisitions Editor: Steve Anglin
   Development Editor: Matthew Moodie
   Technical Reviewer: Jonah Lissner
   Coordinating Editor: Mark Powers
   Copy Editor: Kezia Endsley

Distributed to the book trade worldwide by Springer Science+Business Media New York, 233 Spring Street, 6th Floor, New York, NY 10013. Phone 1-800-SPRINGER, fax (201) 348-4505, e-mail orders-ny@springer-sbm.com, or visit www.springeronline.com. Apress Media, LLC is a California LLC and the sole member (owner) is Springer Science + Business Media Finance Inc (SSBM Finance Inc). SSBM Finance Inc is a **Delaware** corporation.

For information on translations, please e-mail rights@apress.com, or visit http://www.apress.com/rights-permissions.

Apress titles may be purchased in bulk for academic, corporate, or promotional use. eBook versions and licenses are also available for most titles. For more information, reference our Print and eBook Bulk Sales web page at http://www.apress.com/bulk-sales.

Any source code or other supplementary material referenced by the author in this book is available to readers on GitHub via the book's product page, located at www.apress.com/9781484228449. For more detailed information, please visit http://www.apress.com/source-code.

Printed on acid-free paper

Contents at a Glance

Contents

About the Author

Phil Kim, PhD is an experienced MATLAB programmer and user. He also works with algorithms of large datasets drawn from AI, and Machine Learning. He has worked at the Korea Aerospace Research Institute as a Senior Researcher. There, his main task was to develop autonomous flight algorithms and onboard software for unmanned aerial vehicles. He developed an onscreen keyboard program named "Clickey" during his period in the PhD program, which served as a bridge to bring him to his current assignment as a Senior Research Officer at the National Rehabilitation Research Institute of Korea.

About the Technical Reviewer

Jonah Lissner is a research scientist advancing PhD and DSc programs, scholarships, applied projects, and academic journal publications in theoretical physics, power engineering, complex systems, metamaterials, geophysics, and computation theory. He has strong cognitive ability in empiricism and scientific reason for the purpose of hypothesis building, theory learning, and mathematical and axiomatic modeling and testing for abstract problem solving. His dissertations, research publications and projects, CV, journals, blog, novels, and system are listed at `http://Lissnerresearch.weebly.com`.

Acknowledgments

Although I assume that the acknowledgements of most books are not relevant to readers, I would like to offer some words of appreciation, as the following people are very special to me. First, I am deeply grateful to those I studied Deep Learning with at the Modulabs (www.modulabs.co.kr). I owe them for teaching me most of what I know about Deep Learning. In addition, I offer my heartfelt thanks to director S. Kim of Modulabs, who allowed me to work in such a wonderful place from spring to summer. I was able to finish the most of this book at Modulabs.

I also thank president Jeon from Bogonet, Dr. H. You, Dr. Y.S. Kang, and Mr. J. H. Lee from KARI, director S. Kim from Modulabs, and Mr. W. Lee and Mr. S. Hwang from J.MARPLE. They devoted their time and efforts to reading and revising the draft of this book. Although they gave me a hard time throughout the revision process, I finished it without regret.

Lastly, my deepest thanks and love to my wife, who is the best woman I have ever met, and children, who never get bored of me and share precious memories with me.

Introduction

I was lucky enough to witness the world's transition to an information society, followed by a networked environment. I have been living with the changes since I was young. The personal computer opened the door to the world of information, followed by online communication that connected computers via the Internet, and smartphones that connected people. Now, everyone perceives the beginning of the overwhelming wave of artificial intelligence. More and more intelligent services are being introduced, bringing in a new era. *Deep Learning* is the technology that led this wave of intelligence. While it may hand over its throne to other technologies eventually, it stands for now as a cornerstone of this new technology.

Deep Learning is so popular that you can find materials about it virtually anywhere. However, not many of these materials are beginner friendly. I wrote this book hoping that readers can study this subject without the kind of difficulty I experienced when first studying Deep Learning. I also hope that the step-by-step approach of this book can help you avoid the confusion that I faced.

This book is written for two kinds of readers. The first type of reader is one who plans to study Deep Learning in a systematic approach for further research and development. This reader should read all the content from the beginning to end. The example code will be especially helpful for further understanding the concepts. A good deal of effort has been made to construct adequate examples and implement them. The code examples are constructed to be easy to read and understand. They are written in MATLAB for better legibility. There is no better programming language than MATLAB at being able to handle the matrices of Deep Learning in a simple and intuitive manner. The example code uses only basic functions and grammar, so that even those who are not familiar with MATLAB can easily understand the concepts. For those who are familiar with programming, the example code may be easier to understand than the text of this book.

The other kind of reader is one who wants more in-depth information about Deep Learning than what can be obtained from magazines or newspapers, yet doesn't want to study formally. These readers can skip the example code and briefly go over the explanations of the concepts. Such readers may especially want to skip the learning rules of the neural network. In practice, even developers seldom need to implement the learning rules, as various Deep Learning libraries are available. Therefore, those who never need to develop it do not need to bother with it. However, pay closer attention to Chapters 1 and 2 and Chapters 5 and 6. Chapter 6 will be particularly helpful in capturing the most important techniques of Deep Learning, even if you're just reading over the concepts and the results of the examples. Equations occasionally appear to provide a theoretical background. However, they are merely fundamental operations. Simply reading and learning to the point you can tolerate will ultimately lead you to an overall understanding of the concepts.

Organization of the Book

This book consists of six chapters, which can be grouped into three subjects. The first subject is Machine Learning and takes place in Chapter 1. Deep Learning stems from Machine Learning. This implies that if you want to understand the essence of Deep Learning, you have to know the philosophy behind Machine Learning to some extent. Chapter 1 starts with the relationship between Machine Learning and Deep Learning, followed by problem solving strategies and fundamental limitations of Machine Learning. The detailed techniques are not introduced in this chapter. Instead, fundamental concepts that apply to both the neural network and Deep Learning will be covered.

The second subject is the artificial neural network.[1] Chapters 2-4 focus on this subject. As Deep Learning is a type of Machine Learning that employs a neural network, the neural network is inseparable from Deep Learning. Chapter 2 starts with the fundamentals of the neural network: principles of its operation, architecture, and learning rules. It also provides the reason that the simple single-layer architecture evolved to the complex multi-layer architecture. Chapter 3 presents the back-propagation algorithm, which is an important and representative learning rule of the neural network and also employed in Deep Learning. This chapter explains how cost functions and learning rules are related and which cost functions are widely employed in Deep Learning.

Chapter 4 explains how to apply the neural network to classification problems. We have allocated a separate section for classification because it is currently the most prevailing application of Machine Learning. For example, image recognition, one of the primary applications of Deep Learning, is a classification problem.

The third topic is Deep Learning. It is the main topic of this book. Deep Learning is covered in Chapters 5 and 6. Chapter 5 introduces the drivers that enable Deep Learning to yield excellent performance. For a better understanding, it starts with the history of barriers and solutions of Deep Learning. Chapter 6 covers the convolution neural network, which is

---

[1]Unless it can be confused with the neural network of human brain, the artificial neural network is referred to as neural network in this book.

representative of Deep Learning techniques. The convolution neural network is second to none in terms of image recognition. This chapter starts with an introduction of the basic concept and architecture of the convolution neural network as it compares with the previous image recognition algorithms. It is followed by an explanation of the roles and operations of the convolution layer and pooling layer, which act as essential components of the convolution neural network. The chapter concludes with an example of digit image recognition using the convolution neural network and investigates the evolution of the image throughout the layers.

Source Code

All the source code used in this book is available online via the Apress web site at `www.apress.com/9781484228449`. The examples have been tested under MATLAB 2014a. No additional toolbox is required.

CHAPTER 1

■ ■ ■

Machine Learning

You easily find examples where the concepts of Machine Learning and Deep Learning are used interchangeably in the media. However, experts generally distinguish them. If you have decided to study this field, it's important you understand what these words actually mean, and more importantly, how they differ.

What occurred to you when you heard the term "Machine Learning" for the first time? Did you think of something that was similar to Figure 1-1? Then you must admit that you are seriously literal-minded.

Figure 1-1. *Machine Learning or Artificial Intelligence? Courtesy of Euclidean Technologies Management (*www.euclidean.com*)*

Figure 1-1 portrays Artificial Intelligence much more than Machine Learning. Understanding Machine Learning in this way will bring about serious confusion. Although Machine Learning is indeed a branch of Artificial Intelligence, it conveys an idea that is much different from what this image may imply.

© Phil Kim 2017
P. Kim, *MATLAB Deep Learning*, DOI 10.1007/978-1-4842-2845-6_1

In general, Artificial Intelligence, Machine Learning, and Deep Learning are related as follows:

"Deep Learning is a kind of Machine Learning, and
Machine Learning is a kind of Artificial Intelligence."

How is that? It's simple, isn't it? This classification may not be as absolute as the laws of nature, but it is widely accepted.

Let's dig into it a little further. Artificial Intelligence is a very common word that may imply many different things. It may indicate any form of technology that includes some intelligent aspects rather than pinpoint a specific technology field. In contrast, Machine Learning refers to a specific field. In other words, we use Machine Learning to indicate a specific technological group of Artificial Intelligence. Machine Learning itself includes many technologies as well. One of them is Deep Learning, which is the subject of this book.

The fact that Deep Learning is a type of Machine Learning is very important, and that is why we are going through this lengthy review on how Artificial Intelligence, Machine Learning, and Deep Learning are related. Deep Learning has been in the spotlight recently as it has proficiently solved some problems that have challenged Artificial Intelligence. Its performance surely is exceptional in many fields. However, it faces limitations as well. The limitations of Deep Learning stems from its fundamental concepts that has been inherited from its ancestor, Machine Learning. As a type of Machine Learning, Deep Learning cannot avoid the fundamental problems that Machine Learning faces. That is why we need to review Machine Learning before discussing the concept of Deep Learning.

What Is Machine Learning?

In short, Machine Learning is a modeling technique that involves data. This definition may be too short for first-timers to capture what it means. So, let me elaborate on this a little bit. Machine Learning is a technique that figures out the "model" out of "data." Here, the data literally means information such as documents, audio, images, etc. The "model" is the final product of Machine Learning.

Before we go further into the model, let me deviate a bit. Isn't it strange that the definition of Machine Learning only addresses the concepts of data and model and has nothing to do with "learning"? The name itself reflects that the technique analyzes the data and finds the model by itself rather than having a human do it. We call it "learning" because the process resembles being trained with the data to solve the problem of finding a model. Therefore, the data that Machine Learning uses in the modeling process is called "training" data. Figure 1-2 illustrates what happens in the Machine Learning process.

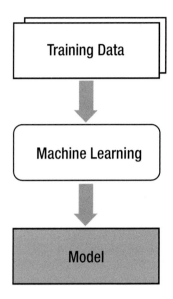

Figure 1-2. *What happens during the machine learning process*

Now, let's resume our discussion about the model. Actually, the model is nothing more than what we want to achieve as the final product. For instance, if we are developing an auto-filtering system to remove spam mail, the spam mail filter is the model that we are talking about. In this sense, we can say the model is what we actually use. Some call the model a *hypothesis*. This term seems more intuitive to those with statistical backgrounds.

Machine Learning is not the only modeling technique. In the field of dynamics, people have been using a certain modeling technique, which employs Newton's laws and describes the motion of objects as a series of equations called equations of motion, for a long time. In the field of Artificial Intelligence, we have the expert system, which is a problem-solving model that is based on the knowledge and know-how of the experts. The model works as well as the experts themselves.

However, there are some areas where laws and logical reasoning are not very useful for modeling. Typical problems can be found where intelligence is involved, such as image recognition, speech recognition, and natural language processing. Let me give you an example. Look at Figure 1-3 and identify the numbers.

Figure 1-3. *How does a computer identify numbers when they have no recognizable pattern?*

I'm sure you have completed the task in no time. Most people do. Now, let's make a computer do the same thing. What do we do? If we use a traditional modeling technique, we will need to find some rule or algorithm to distinguish the written numbers. Hmm, why don't we apply the rules that you have just used to identify the numbers in your brain? Easy enough, isn't it? Well, not really. In fact, this is a very challenging problem. There was a time when researchers thought it must be a piece of cake for computers to do this, as it is very easy for even a human and computers are able to calculate much faster than humans. Well, it did not take very long until they realized their misjudgment.

How were you able to identify the numbers without a clear specification or a rule? It is hard to answer, isn't it? But, why? It is because we have never learned such a specification. From a young age, we have just learned that this is 0, and that this is 1. We just thought that's what it is and became better at distinguishing numbers as we faced a variety of numbers. Am I right?

What about computers, then? Why don't we let computers do the same thing? That's it! Congratulations! You have just grasped the concept of Machine Learning. Machine Learning has been created to solve the problems for which analytical models are hardly available. The primary idea of Machine Learning is to achieve a model using the training data when equations and laws are not promising.

Challenges with Machine Learning

We just discovered that Machine Learning is the technique used to find (or learn) a model from the data. It is suitable for problems that involve intelligence, such as image recognition and speech recognition, where physical laws or mathematical equations fail to produce a model. On the one hand, the approach that Machine Learning uses is what makes the process work. On the other hand, it brings inevitable problems. This section provides the fundamental issues Machine Learning faces.

Once the Machine Learning process finds the model from the training data, we apply the model to the actual field data. This process is illustrated in Figure 1-4. The vertical flow of the figure indicates the learning process, and the trained model is described as the horizontal flow, which is called inference.

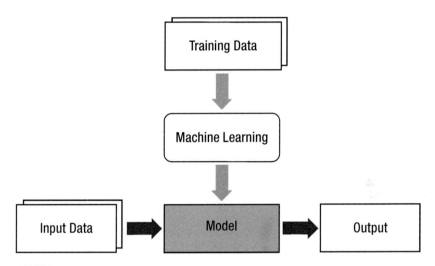

Figure 1-4. *Applying a model based on field data*

The data that is used for modeling in Machine Learning and the data supplied in the field application are distinct. Let's add another block to this image, as shown in Figure 1-5, to better illustrate this situation.

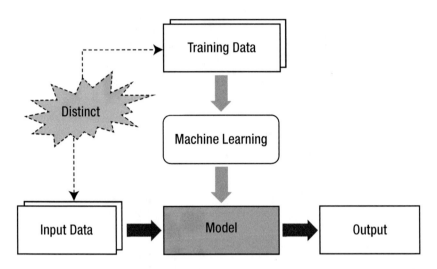

Figure 1-5. *Training and input data are sometimes very distinct*

The distinctness of the training data and input data is the structural challenge that Machine Learning faces. It is no exaggeration to say that every problem of Machine Learning originates from this. For example, what about using training data, which is composed of handwritten notes from a single person? Will the model successfully recognize the other person's handwriting? The possibility will be very low.

No Machine Learning approach can achieve the desired goal with the wrong training data. The same ideology applies to Deep Learning. Therefore, it is critical for Machine Learning approaches to obtain unbiased training data that adequately reflects the characteristics of the field data. The process used to make the model performance consistent regardless of the training data or the input data is called *generalization*. The success of Machine Learning relies heavily on how well the generalization is accomplished.

Overfitting

One of the primary causes of corruption of the generalization process is *overfitting*. Yes, another new term. However, there is no need to be frustrated. It is not a new concept at all. It will be much easier to understand with a case study than with just sentences.

Consider a classification problem shown in Figure 1-6. We need to divide the position (or coordinate) data into two groups. The points on the figure are the training data. The objective is to determine a curve that defines the border of the two groups using the training data.

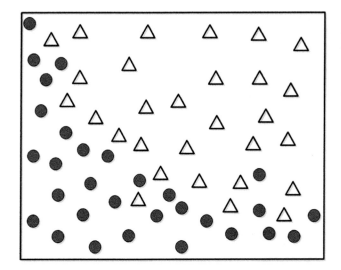

Figure 1-6. *Determine a curve to divide two groups of data*

Although we see some outliers that deviate from the adequate area, the curve shown in Figure 1-7 seems to act as a reasonable border between the groups.

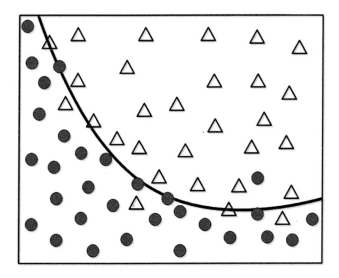

Figure 1-7. *Curve to differentiate between two types of data*

When we judge this curve, there are some points that are not correctly classified according to the border. What about perfectly grouping the points using a complex curve, as shown in Figure 1-8?

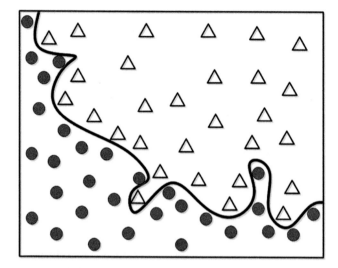

Figure 1-8. *Better grouping, but at what cost?*

This model yields the perfect grouping performance for the training data. How does it look? Do you like this model better? Does it seem to reflect correctly the general behavior?

Now, let's use this model in the real world. The new input to the model is indicated using the symbol ■, as shown in Figure 1-9.

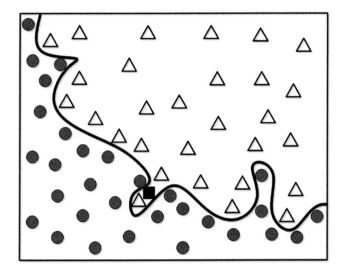

Figure 1-9. *The new input is placed into the data*

This proud error-free model identifies the new data as a class Δ. However, the general trend of the training data tells us that this is doubtable. Grouping it as a class ● seems more reasonable. What happened to the model that yielded 100% accuracy for the training data?

Let's take another look at the data points. Some outliers penetrate the area of the other group and disturb the boundary. In other words, this data contains much noise. The problem is that there is no way for Machine Learning to distinguish this. As Machine Learning considers all the data, even the noise, it ends up producing an improper model (a curve in this case). This would be penny-wise and pound-foolish. As you may notice here, the training data is not perfect and may contain varying amounts of noise. If you believe that every element of the training data is correct and fits the model precisely, you will get a model with lower generalizability. This is called *overfitting*.

Certainly, because of its nature, Machine Learning should make every effort to derive an excellent model from the training data. However, a working model of the training data may not reflect the field data properly. This does not mean that we should make the model less accurate than the training data on purpose. This will undermine the fundamental strategy of Machine Learning.

Now we face a dilemma—reducing the error of the training data leads to overfitting that degrades generalizability. What do we do? We confront it, of course! The next section introduces the techniques that prevent overfitting.

Confronting Overfitting

Overfitting significantly affects the level of performance of Machine Learning. We can tell who is a pro and who is an amateur by watching their respective approaches in dealing with overfitting. This section introduces two typical methods used to confront overfitting: regularization and validation.

Regularization is a numerical method that attempts to construct a model structure as simple as possible. The simplified model can avoid the effects of overfitting at the small cost of performance. The grouping problem of the previous section can be used as a good example. The complex model (or curve) tends to be overfitting. In contrast, although it fails to classify correctly some points, the simple curve reflects the overall characteristics of the group much better. If you understand how it works, that is enough for now. We will revisit regularization with further details in Chapter Three's "Cost Function and Learning Rule" section.

We are able to tell that the grouping model is overfitted because the training data is simple, and the model can be easily visualized. However, this is not the case for most situations, as the data has higher dimensions. We cannot draw the model and intuitively evaluate the effects of overfitting for such data. Therefore, we need another method to determine whether the trained model is overfitted or not. This is where *validation* comes into play.

The validation is a process that reserves a part of the training data and uses it to monitor the performance. The validation set is not used for the training process. Because the modeling error of the training data fails to indicate overfitting, we use some of the training data to check if the model is overfitted. We can say that the model is overfitted when the trained model yields a low level of performance to the reserved data input. In this case, we will modify the model to prevent the overfitting. Figure 1-10 illustrates the division of the training data for the validation process.

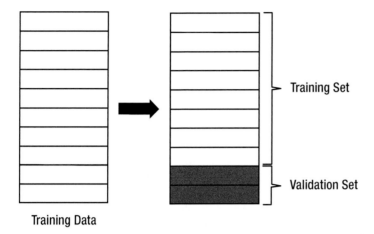

Training Data

Figure 1-10. *Dividing the training data for the validation process*

When validation is involved, the training process of Machine Learning proceeds by the following steps:

1. Divide the training data into two groups: one for training and the other for validation. As a rule of thumb, the ratio of the training set to the validation set is 8:2.

2. Train the model with the training set.

3. Evaluate the performance of the model using the validation set.

    a. If the model yields satisfactory performance, finish the training.

    b. If the performance does not produce sufficient results, modify the model and repeat the process from Step 2.

Cross-validation is a slight variation of the validation process. It still divides the training data into groups for the training and validation, but keeps changing the datasets. Instead of retaining the initially divided sets, cross-validation repeats the division of the data. The reason for doing this is that the model can be overfitted even to the validation set when it is fixed. As the cross-validation maintains the randomness of the validation dataset, it can better detect the overfitting of the model. Figure 1-11 describes the concept of cross-validation. The dark shades indicate the validation data, which is randomly selected throughout the training process.

11

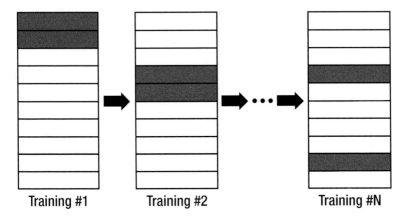

Figure 1-11. *Cross-validation*

Types of Machine Learning

Many different types of Machine Learning techniques have been developed to solve problems in various fields. These Machine Learning techniques can be classified into three types depending on the training method (see Figure 1-12).

- Supervised learning

- Unsupervised learning

- Reinforcement learning

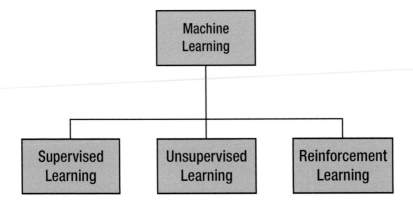

Figure 1-12. *Three types of Machine Learning techniques*

Supervised learning is very similar to the process in which a human learns things. Consider that humans obtain new knowledge as we solve exercise problems.

1. Select an exercise problem. Apply current knowledge to solve the problem. Compare the answer with the solution.

2. If the answer is wrong, modify current knowledge.

3. Repeat Steps 1 and 2 for all the exercise problems.

When we apply an analogy between this example and the Machine Learning process, the exercise problems and solutions correspond to the training data, and the knowledge corresponds to the model. The important thing is the fact that we need the solutions. This is the vital aspect of the supervised learning. Its name even implies the tutoring in which the teacher gives solutions to the students to memorize.

In supervised learning, each training dataset should consist of input and correct output pairs. The correct output is what the model is supposed to produce for the given input.

```
{ input, correct output }
```

Learning in supervised learning is the series of revisions of a model to reduce the difference between the correct output and the output from the model for the same input. If a model is perfectly trained, it will produce a correct output that corresponds to the input from the training data.

In contrast, the training data of the unsupervised learning contains only inputs without correct outputs.

```
{ input }
```

At a first glance, it may seem difficult to understand how to train without correct outputs. However, many methods of this type have been developed already. Unsupervised learning is generally used for investigating the characteristics of the data and preprocessing the data. This concept is similar to a student who just sorts out the problems by construction and attribute and doesn't learn how to solve them because there are no known correct outputs.

Reinforcement learning employs sets of input, some output, and grade as training data. It is generally used when optimal interaction is required, such as control and game plays.

```
{ input, some output, grade for this output }
```

This book only covers supervised learning. It is used for more applications compared to unsupervised learning and reinforcement learning, and more importantly, it is the first concept you will study when entering the world of Machine Learning and Deep Learning.

Classification and Regression

The two most common types of application of supervised learning are classification and regression. These words may sound unfamiliar, but are actually not so challenging.

Let's start with classification. This may be the most prevailing application of Machine Learning. The classification problem focuses on literally finding the classes to which the data belongs. Some examples may help.

Spam mail filtering service ➔ Classifies the mails by regular or spam

Digit recognition service ➔ Classifies the digit image into one of 0-9

Face recognition service ➔ Classifies the face image into one of the registered users

We addressed in the previous section that supervised learning requires input and correct output pairs for the training data. Similarly, the training data of the classification problem looks like this:

```
{ input, class }
```

In the classification problem, we want to know which class the input belongs to. So the data pair has the class in place of the correct output corresponding to the input.

Let's proceed with an example. Consider the same grouping problem that we have been discussing. The model we want Machine Learning to answer is which one of the two classes (Δ and ●) does the user's input coordinate (X, Y) belong (see Figure 1-13).

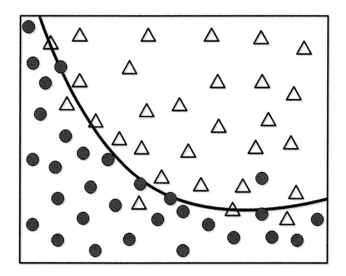

Figure 1-13. *Same data viewed from the perspective of classification*

In this case, the training data of N sets of the element will look like Figure 1-14.

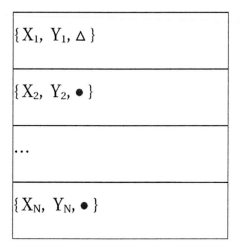

Figure 1-14. *Classifying the data*

In contrast, the regression does not determine the class. Instead, it estimates a value. As an example, if you have datasets of age and income (indicated with a ●) and want to find the model that estimates income by age, it becomes a regression problem (see Figure 1-15).[1]

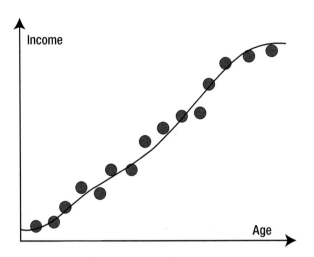

Figure 1-15. *Datasets of age and income*

The dataset of this example will look like the table in Figure 1-16, where X and Y are age and income, respectively.

---

[1]The original meaning of the word "regress" is to go back to an average. Francis Galton, a British geneticist, researched the correlation of the height of parents and children and found out that the individual height converged to the average of the total population. He named his methodology "regression analysis."

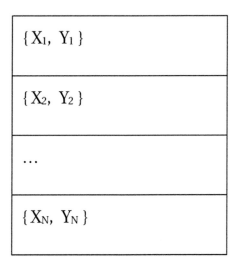

Figure 1-16. *Classifying the age and income data*

Both classification and regression are parts of supervised learning. Therefore, their training data is equally in the form of {input, correct output}. The only difference is the type of correct outputs—classification employs classes, while the regression requires values.

In summary, analysis can become classification when it needs a model to judge which group the input data belongs to and regression when the model estimates the trend of the data.

Just for reference, one of the representative applications of unsupervised learning is *clustering*. It investigates the characteristics of the individual data and categorizes the related data. It is very easy to confuse clustering and classification, as their results are similar. Although they yield similar outputs, they are two completely different approaches. We have to keep in mind that clustering and classification are distinct terms. When you encounter the term *clustering*, just remind yourself that it focuses on unsupervised learning.

Summary

Let's briefly recap what we covered in this chapter:

- Artificial Intelligence, Machine Learning, and Deep Learning are distinct. But they are related to each other in the following way: "Deep Learning is a kind of Machine Learning, and Machine Learning is a kind of Artificial Intelligence".

- Machine Learning is an inductive approach that derives a model from the training data. It is useful for image recognition, speech recognition, and natural language processing etc.

- The success of Machine Learning heavily relies on how well the generalization process is implemented. In order to prevent performance degradation due to the differences between the training data and actual input data, we need a sufficient amount of unbiased training data.

- Overfitting occurs when the model has been overly customized to the training data that it yields poor performance for the actual input data, while its performance for the training data is excellent. Overfitting is one of the primary factors that reduces the generalization performance.

- Regularization and validation are the typical approaches used to solve the overfitting problem. Regularization is a numerical method that yields the simplest model as possible. In contrast, validation tries to detect signs of overfitting during training and takes action to prevent it. A variation of validation is cross-validation.

- Depending on the training method, Machine Learning can be supervised learning, unsupervised learning, and reinforcement learning. The formats of the training data for theses learning methods are shown here.

Training Method	Training Data
Supervised Learning	`{ input, correct output }`
Unsupervised Learning	`{ input }`
Reinforced Learning	`{ input, some output, grade for this output }`

- Supervised learning can be divided into classification and regression, depending on the usage of the model. Classification determines which group the input data belongs to. The correct output of the classification is given as categories. In contrast, regression predicts values and takes the values for the correct output in the training data.

CHAPTER 2

■ ■ ■

Neural Network

This chapter introduces the neural network, which is widely used as the model for Machine Learning. The neural network has a long history of development and a vast amount of achievement from research works. There are many books available that purely focus on the neural network. Along with the recent growth in interest for Deep Learning, the importance of the neural network has increased significantly as well. We will briefly review the relevant and practical techniques to better understand Deep Learning. For those who are new to the concept of the neural network, we start with the fundamentals.

First, we will see how the neural network is related to Machine Learning. The models of Machine Learning can be implemented in various forms. The neural network is one of them. Simple isn't it? Figure 2-1 illustrates the relationship between Machine Learning and the neural network. Note that we have the neural network in place of the model, and the learning rule in place of Machine Learning. In context of the neural network, the process of determining the model (neural network) is called the learning rule. This chapter explains the learning rules for a single-layer neural network. The learning rules for a multi-layer neural network are addressed in Chapter 3.

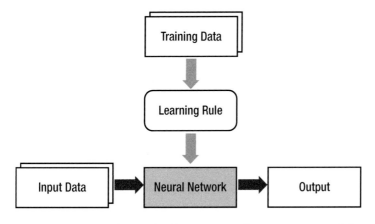

Figure 2-1. *The relationship between Machine Learning and the neural network*

© Phil Kim 2017

P. Kim, *MATLAB Deep Learning*, DOI 10.1007/978-1-4842-2845-6_2

Nodes of a Neural Network

Whenever we learn something, our brain stores the knowledge. The computer uses memory to store information. Although they both store information, their mechanisms are very different. The computer stores information at specified locations of the memory, while the brain alters the association of neurons. The neuron itself has no storage capability; it just transmits signals from one neurons to the other. The brain is a gigantic network of these neurons, and the association of the neurons forms specific information.

The neural network imitates the mechanism of the brain. As the brain is composed of connections of numerous neurons, the neural network is constructed with connections of nodes, which are elements that correspond to the neurons of the brain. The neural network mimics the neurons' association, which is the most important mechanism of the brain, using the weight value. The following table summarizes the analogy between the brain and neural network.

Brain	Neural Network
Neuron	Node
Connection of neurons	Connection weight

Explaining this any further using text may cause more confusion. Look at a simple example for a better understanding of the neural network's mechanism. Consider a node that receives three inputs, as shown in Figure 2-2.

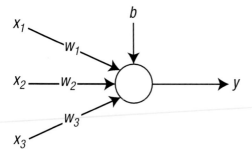

Figure 2-2. *A node that receives three inputs*

The circle and arrow of the figure denote the node and signal flow, respectively. x_1, x_2, and x_3 are the input signals. w_1, w_2, and w_3 are the weights for the corresponding signals. Lastly, b is the bias, which is another factor associated with the storage of information. In other words, the information of the neural net is stored in the form of weights and bias.

The input signal from the outside is multiplied by the weight before it reaches the node. Once the weighted signals are collected at the node, these values are added to be the weighted sum. The weighted sum of this example is calculated as follows:

$$v = (w_1 \times x_1) + (w_2 \times x_2) + (w_3 \times x_3) + b$$

This equation indicates that the signal with a greater weight has a greater effect. For instance, if the weight w_1 is 1, and w_2 is 5, then the signal x_2 has five times larger effect than that of x_1. When w_1 is zero, x_1 is not transmitted to the node at all. This means that x_1 is disconnected from the node. This example shows that the weights of the neural network imitate how the brain alters the association of the neurons.

The equation of the weighted sum can be written with matrices as:

$$v = wx + b$$

where w and x are defined as:

$$w = \begin{bmatrix} w_1 & w_2 & w_3 \end{bmatrix} \qquad x = \begin{bmatrix} x_1 \\ x_2 \\ x_3 \end{bmatrix}$$

Finally, the node enters the weighted sum into the activation function and yields its output. The activation function determines the behavior of the node.

$$y = \varphi(v)$$

$\varphi(\cdot)$ of this equation is the activation function. Many types of activation functions are available in the neural network. We will elaborate on them later.

Let's briefly review the mechanism of the neural net. The following process is conducted inside the neural net node:

1. The weighted sum of the input signals is calculated.

$$v = w_1 x_1 + w_2 x_2 + w_3 x_3 + b$$
$$= wx + b$$

2. The output from the activation function to the weighted sum is passed outside.

$$y = \varphi(v) = \varphi(wx + b)$$

Layers of Neural Network

As the brain is a gigantic network of the neurons, the neural network is a network of nodes. A variety of neural networks can be created depending on how the nodes are connected. One of the most commonly used neural network types employs a layered structure of nodes as shown in Figure 2-3.

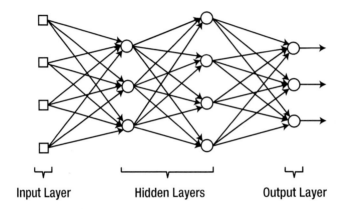

Input Layer   Hidden Layers   Output Layer

Figure 2-3. *A layered structure of nodes*

The group of square nodes in Figure 2-3 is called the input layer. The nodes of the input layer merely act as the passage that transmits the input signals to the next nodes. Therefore, they do not calculate the weighted sum and activation function. This is the reason that they are indicated by squares and distinguished from the other circular nodes. In contrast, the group of the rightmost nodes is called the output layer. The output from these nodes becomes the final result of the neural network. The layers in between the input and output layers are called *hidden layers*. They are given this name because they are not accessible from the outside of the neural network.

The neural network has been developed from a simple architecture to a more and more complex structure. Initially, neural network pioneers had a very simple architecture with only input and output layers, which are called *single-layer neural networks*. When hidden layers are added to a single-layer neural network, this produces a multi-layer neural network. Therefore, the multi-layer neural network consists of an input layer, hidden layer(s), and output layer. The neural network that has a single hidden layer is called a *shallow neural network* or a vanilla neural network. A multi-layer neural network that contains two or more hidden layers is called a *deep neural network*. Most of the contemporary neural networks used in practical applications are deep neural networks. The following table summarizes the branches of the neural network depending on the layer architecture.

Single-Layer Neural Network		Input Layer – Output Layer
Multi-Layer Neural Network	Shallow Neural Network	Input Layer – Hidden Layer – Output Layer
	Deep Neural Network	Input Layer – Hidden Layers – Output Layers

The reason that we classify the multi-layer neural network by these two types has to do with its historical background of development. The neural network started as the single-layer neural network and evolved to the shallow neural network, followed by the deep neural network. The deep neural network has not been seriously highlighted until the mid-2000s, after two decades had passed since the development of the shallow neural network. Therefore, for a long time, the multi-layer neural network meant just the single hidden-layer neural network. When the need to distinguish multiple hidden layers arose, they gave a separate name to the deep neural network. See Figure 2-4.

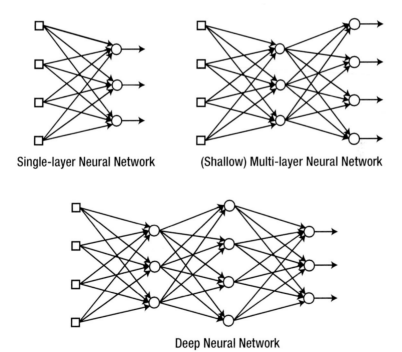

Single-layer Neural Network   (Shallow) Multi-layer Neural Network

Deep Neural Network

Figure 2-4. *The branches of the neural network depend on the layer architecture*

In the layered neural network, the signal enters the input layer, passes through the hidden layers, and leaves through the output layer. During this process, the signal advances layer by layer. In other words, the nodes on one layer receive the signal simultaneously and send the processed signal to the next layer at the same time.

Let's follow a simple example to see how the input data is processed as it passes through the layers. Consider the neural network with a single hidden layer shown in Figure 2-5.

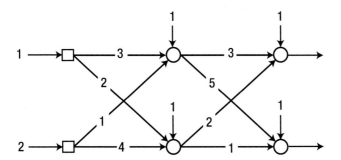

Figure 2-5. *A neural network with a single hidden layer*

Just for convenience, the activation function of each node is assumed to be a linear function shown in Figure 2-6. This function allows the nodes to send out the weighted sum itself.

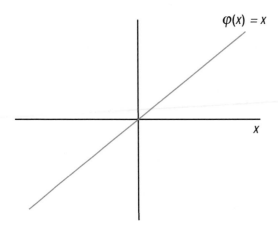

Figure 2-6. *The activation function of each node is a linear function*

Now we will calculate the output from the hidden layer (Figure 2-7). As previously addressed, no calculation is needed for the input nodes, as they just transmit the signal.

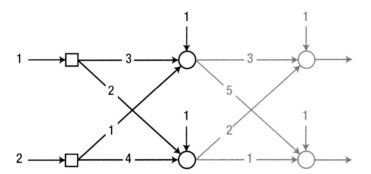

Figure 2-7. *Calculate the output from the hidden layer*

The first node of the hidden layer calculates the output as:

Weighted sum: $v = (3{\times}1) + (1{\times}2) + 1 = 6$

Output: $y = \varphi(v) = v = 6$

In a similar manner, the second node of the hidden layer calculates the output as:

Weighted sum: $v = (2{\times}1) + (4{\times}2) + 1 = 11$

Output: $y = \varphi(v) = v = 11$

The weighted sum calculations can be combined in a matrix equation as follows:

$$v = \begin{bmatrix} 3{\times}1+1{\times}2+1 \\ 2{\times}1+4{\times}2+1 \end{bmatrix} = \begin{bmatrix} 3 & 1 \\ 2 & 4 \end{bmatrix}\begin{bmatrix} 1 \\ 2 \end{bmatrix} + \begin{bmatrix} 1 \\ 1 \end{bmatrix} = \begin{bmatrix} 6 \\ 11 \end{bmatrix}$$

The weights of the first node of the hidden layer lay in the first row, and the weights of the second node are in the second row. This result can be generalized as the following equation:

$$v = Wx + b \qquad\qquad \text{(Equation 2.1)}$$

where *x* is the input signal vector and *b* is the bias vector of the node. The matrix *W* contains the weights of the hidden layer nodes on the corresponding rows. For the example neural network, *W* is given as:

$$W = \begin{bmatrix} -- \text{ weights of the first node } -- \\ -- \text{ weights of the second node } -- \end{bmatrix} = \begin{bmatrix} 3 & 1 \\ 2 & 4 \end{bmatrix}$$

Since we have all the outputs from the hidden layer nodes, we can determine the outputs of the next layer, which is the output layer. Everything is identical to the previous calculation, except that the input signal comes from the hidden layer.

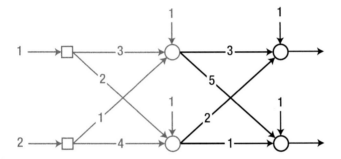

Figure 2-8. *Determine the outputs of the output layer*

Let's use the matrix form of Equation 2.1 to calculate the output.

$$\text{Weighted sum: } v = \begin{bmatrix} 3 & 2 \\ 5 & 1 \end{bmatrix} \begin{bmatrix} 6 \\ 11 \end{bmatrix} + \begin{bmatrix} 1 \\ 1 \end{bmatrix} = \begin{bmatrix} 41 \\ 42 \end{bmatrix}$$

$$\text{Output: } y = \varphi(v) = v = \begin{bmatrix} 41 \\ 42 \end{bmatrix}$$

How was that? The process may be somewhat cumbersome, but there is nothing difficult in the calculation itself. As we just saw, the neural network is nothing more than a network of layered nodes, which performs only simple calculations. It does not involve any difficult equations or a complicated architecture. Although it appears to be simple, the neural network has been breaking all performance records for the major Machine Learning fields, such as image recognition and speech recognition. Isn't it interesting? It seems like the quote, "All the truth is simple" is an apt description.

I must leave a final comment before wrapping up the section. We used a linear equation for the activation of the hidden nodes, just for convenience. This is not practically correct. The use of a linear function for the nodes negates the effect of adding a layer. In this case, the model is mathematically identical to a single-layer neural network, which does not have hidden layers. Let's see what really happens. Substituting the equation of weighted sum of the hidden layer into the equation of weighted sum of the output layer yields the following equation:

$$
\begin{aligned}
v &= \begin{bmatrix} 3 & 2 \\ 5 & 1 \end{bmatrix}\begin{bmatrix} 6 \\ 11 \end{bmatrix} + \begin{bmatrix} 1 \\ 1 \end{bmatrix} \\
&= \begin{bmatrix} 3 & 2 \\ 5 & 1 \end{bmatrix}\left(\begin{bmatrix} 3 & 1 \\ 2 & 4 \end{bmatrix}\begin{bmatrix} 1 \\ 2 \end{bmatrix} + \begin{bmatrix} 1 \\ 1 \end{bmatrix}\right) + \begin{bmatrix} 1 \\ 1 \end{bmatrix} \\
&= \begin{bmatrix} 3 & 2 \\ 5 & 1 \end{bmatrix}\begin{bmatrix} 3 & 1 \\ 2 & 4 \end{bmatrix}\begin{bmatrix} 1 \\ 2 \end{bmatrix} + \begin{bmatrix} 3 & 2 \\ 5 & 1 \end{bmatrix}\begin{bmatrix} 1 \\ 1 \end{bmatrix} + \begin{bmatrix} 1 \\ 1 \end{bmatrix} \\
&= \begin{bmatrix} 13 & 11 \\ 17 & 9 \end{bmatrix}\begin{bmatrix} 1 \\ 2 \end{bmatrix} + \begin{bmatrix} 6 \\ 7 \end{bmatrix}
\end{aligned}
$$

This matrix equation indicates that this example neural network is equivalent to a single layer neural network as shown in Figure 2-9.

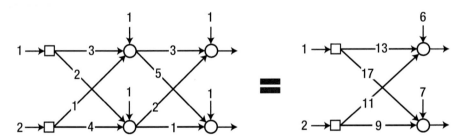

Figure 2-9. *This example neural network is equivalent to a single layer neural network*

Keep in mind that the hidden layer becomes ineffective when the hidden nodes have linear activation functions. However, the output nodes may, and sometimes must, employ linear activation functions.

Supervised Learning of a Neural Network

This section introduces the concepts and process of supervised learning of the neural network. It is addressed in the "Types of Machine Learning" section in Chapter 1. Of the many training methods, this book covers only supervised learning. Therefore, only supervised learning is discussed for the neural network as well. In the big picture, supervised learning of the neural network proceeds in the following steps:

1. Initialize the weights with adequate values.

2. Take the "input" from the training data, which is formatted as { input, correct output }, and enter it into the neural network. Obtain the output from the neural network and calculate the error from the correct output.

3. Adjust the weights to reduce the error.

4. Repeat Steps 2-3 for all training data

These steps are basically identical to the supervised learning process of the "Types of Machine Learning" section. This makes sense because the training of supervised learning is a process that modifies the model to reduce the difference between the correct output and model's output. The only difference is that the modification of the model becomes the changes in weights for the neural network. Figure 2-10 illustrates the concept of supervised learning that has been explained so far. This will help you clearly understand the steps described previously.

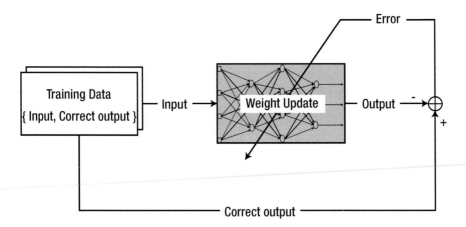

Figure 2-10. *The concept of supervised learning*

Training of a Single-Layer Neural Network: Delta Rule

As previously addressed, the neural network stores information in the form of weights.[1] Therefore, in order to train the neural network with new information, the weights should be changed accordingly. The systematic approach to modifying the weights according to the given information is called the learning rule. Since training is the only way for the neural network to store the information systematically, the learning rule is a vital component in neural network research.

In this section, we deal with the *delta rule*,[2] the representative learning rule of the single-layer neural network. Although it is not capable of multi-layer neural network training, it is very useful for studying the important concepts of the learning rule of the neural network.

Consider a single-layer neural network, as shown in Figure 2-11. In the figure, d_i is the correct output of the output node i.

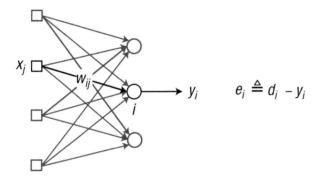

$$e_i \triangleq d_i - y_i$$

Figure 2-11. A single-layer neural network

Long story short, the delta rule adjusts the weight as the following algorithm:

> "If an input node contributes to the error of the output node, the weight between the two nodes is adjusted in proportion to the input value, x_j and the output error, e_i."

---

[1] Unless otherwise noticed, the weight in this book includes bias as well.
[2] It is also referred to as Adaline rule as well as Widrow-Hoff rule.

This rule can be expressed in equation as:

$$w_{ij} \leftarrow w_{ij} + \alpha e_i x_j \qquad \text{(Equation 2.2)}$$

where

x_j = The output from the input node j, (j = 1, 2, 3)

e_i = The error of the output node i

w_{ij} = The weight between the output node i and input node j

α = Learning rate ($0 < \alpha \leq 1$)

The learning rate, α, determines how much the weight is changed per time. If this value is too high, the output wanders around the solution and fails to converge. In contrast, if it is too low, the calculation reaches the solution too slowly.

To take a concrete example, consider the single-layer neural network, which consists of three input nodes and one output node, as shown in Figure 2-12. For convenience, we assume no bias for the output node at this time. We use a linear activation function; i.e., the weighted sum is directly transferred to the output.

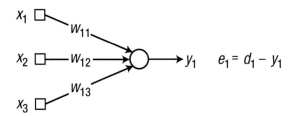

Figure 2-12. *A single-layer neural network with three input nodes and one output node*

Note that the first number of the subscript (1) indicates the node number to which the input enters. For example, the weight between the input node 2 and output node 1 is denoted as w_{12}. This notation enables an easier matrix operation; the weights associated with the node i are allocated at the i-th row of the weight matrix.

Applying the delta rule of Equation 2.2 to the example neural network yields the renewal of the weights as:

$$w_{11} \leftarrow w_{11} + \alpha e_1 x_1$$
$$w_{12} \leftarrow w_{12} + \alpha e_1 x_2$$
$$w_{13} \leftarrow w_{13} + \alpha e_1 x_3$$

Let's summarize the training process using the delta rule for the single-layer neural network.

1. Initialize the weights at adequate values.

2. Take the "input" from the training data of { input, correct output } and enter it to the neural network. Calculate the error of the output, y_i, from the correct output, d_i, to the input.

$$e_i = d_i - y_i$$

3. Calculate the weight updates according to the following delta rule:

$$\Delta w_{ij} = \alpha\, e_i\, x_j$$

4. Adjust the weights as:

$$w_{ij} \leftarrow w_{ij} + \Delta w_{ij}$$

5. Perform Steps 2-4 for all training data.

6. Repeat Steps 2-5 until the error reaches an acceptable tolerance level.

These steps are almost identical to that of the process for the supervised learning in the "Supervised Learning of a Neural Network" section. The only difference is the addition of Step 6. Step 6 just states that the whole training process is repeated. Once Step 5 has been completed, the model has been trained with every data point. Then, why do we train it using all of the same training data? This is because the delta rule searches for the solution as it repeats the process, rather than solving it all at once.[3] The whole process repeats, as retraining the model with the same data may improve the model.

Just for reference, the number of training iterations, in each of which all training data goes through Steps 2-5 once, is called an *epoch*. For instance, epoch = 10 means that the neural network goes through 10 repeated training processes with the same dataset.

---

[3]The delta rule is a type of numerical method called *gradient descent*. The gradient descent starts from the initial value and proceeds to the solution. Its name originates from its behavior whereby it searches for the solution as if a ball rolls down the hill along the steepest path. In this analogy, the position of the ball is the occasional output from the model, and the bottom is the solution. It is noteworthy that the gradient descent method cannot drop the ball to the bottom with just one throw.

Are you able to follow this section so far? Then you have learned most of the key concepts of the neural network training. Although the equations may vary depending on the learning rule, the essential concepts are relatively the same. Figure 2-13 illustrates the training process described in this section.

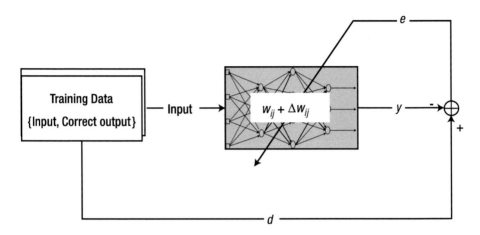

Figure 2-13. *The training process*

Generalized Delta Rule

This section touches on some theoretical aspects of the delta rule. However, you don't need to be frustrated. We will go through the most essential subjects without elaborating too much on the specifics.

The delta rule of the previous section is rather obsolete. Later studies have uncovered that there exists a more generalized form of the delta rule. For an arbitrary activation function, the delta rule is expressed as the following equation.

$$w_{ij} \leftarrow w_{ij} + \alpha \delta_i x_j \qquad \text{(Equation 2.3)}$$

It is the same as the delta rule of the previous section, except that e_i is replaced with δ_i. In this equation, δ_i is defined as:

$$\delta_i = \varphi'(v_i)e_i \qquad \text{(Equation 2.4)}$$

where

e_i = The error of the output node i

v_i = The weighted sum of the output node i

φ' = The derivative of the activation function φ of the output node i

Recall that we used a linear activation function of $\varphi(x) = x$ for the example. The derivative of this function is $\varphi'(x) = 1$. Substituting this value into Equation 2.4 yields δ_i as:

$$\delta_i = e_i$$

Plugging this equation into Equation 2.3 results in the same formula as the delta rule in Equation 2.2. This fact indicates that the delta rule in Equation 2.2 is only valid for linear activation functions.

Now, we can derive the delta rule with the sigmoid function, which is widely used as an activation function. The sigmoid function is defined as shown in Figure 2-14.[4]

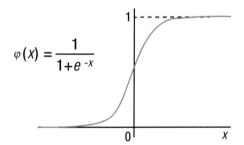

Figure 2-14. *The sigmoid function defined*

We need the derivative of this function, which is given as:

$$\varphi'(x) = \varphi(x)\big(1-\varphi(x)\big)$$

Substituting this derivative into Equation 2.4 yields δ_i as:

$$\delta_i = \varphi'(v_i)e_i = \varphi(v_i)\big(1-\varphi(v_i)\big)e_i$$

_____

[4]The output from a sigmoid function is within the range of 0-1. This behavior of the sigmoid function is useful when the neural network produces probability outputs.

Again, plugging this equation into Equation 2.3 gives the delta rule for the sigmoid function as:

$$w_{ij} \leftarrow w_{ij} + \alpha\, \varphi(v_i)\big(1-\varphi(v_i)\big)e_i x_j \qquad \text{(Equation 2.5)}$$

Although the weight update formula is rather complicated, it maintains the identical fundamental concept where the weight is determined in proportion to the output node error, e_i and the input node value, x_j.

SGD, Batch, and Mini Batch

The schemes that are used to calculate the weight update, Δw_{ij}, are introduced in this section. Three typical schemes are available for supervised learning of the neural network.

Stochastic Gradient Descent

The Stochastic Gradient Descent (SGD) calculates the error for each training data and adjusts the weights immediately. If we have 100 training data points, the SGD adjusts the weights 100 times. Figure 2-15 shows how the weight update of the SGD is related to the entire training data.

Weight Update → Training

Training Data

Figure 2-15. *How the weight update of the SGD is related to the entire training data*

As the SGD adjusts the weight for each data point, the performance of the neural network is crooked while the undergoing the training process. The name "stochastic" implies the random behavior of the training process. The SGD calculates the weight updates as:

$$\Delta w_{ij} = \alpha \delta_i x_j$$

This equation implies that all the delta rules of the previous sections are based on the SGD approach.

Batch

In the batch method, each weight update is calculated for all errors of the training data, and the average of the weight updates is used for adjusting the weights. This method uses all of the training data and updates only once. Figure 2-16 explains the weight update calculation and training process of the batch method.

Average of Weight Updates
→ Training

Training Data

Figure 2-16. *The batch method's weight update calculation and training process*

The batch method calculates the weight update as:

$$\Delta w_{ij} = \frac{1}{N}\sum_{k=1}^{N}\Delta w_{ij}(k) \qquad \text{(Equation 2.6)}$$

where $\Delta w_{ij}(k)$ is the weight update for the k-th training data and N is the total number of the training data.

Because of the averaged weight update calculation, the batch method consumes a significant amount of time for training.

Mini Batch

The mini batch method is a blend of the SGD and batch methods. It selects a part of the training dataset and uses them for training in the batch method. Therefore, it calculates the weight updates of the selected data and trains the neural network with the averaged weight update. For example, if 20 arbitrary data points are selected out of 100 training data points, the batch method is applied to the 20 data points. In this case, a total of five weight adjustments are performed to complete the training process for all the data points (5 = 100/20). Figure shows 2-17 how the mini batch scheme selects training data and calculates the weight update.

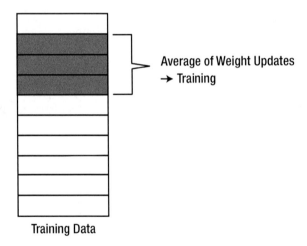

Figure 2-17. How the mini batch scheme selects training data and calculates the weight update

The mini batch method, when it selects an appropriate number of data points, obtains the benefits from both methods: speed from the SGD and stability from the batch. For this reason, it is often utilized in Deep Learning, which manipulates a significant amount of data.

Now, let's delve a bit into the SGD, batch, and mini batch in terms of the epoch. The epoch is briefly introduced in the "Training of a Single-Layer Neural Network: Delta Rule" section. As a recap, the epoch is the number of completed training cycles for all of the training data. In the batch method, the number of training cycles of the neural network equals an epoch, as shown in Figure 2-18. This makes perfect sense because the batch method utilizes all of the data for one training process.

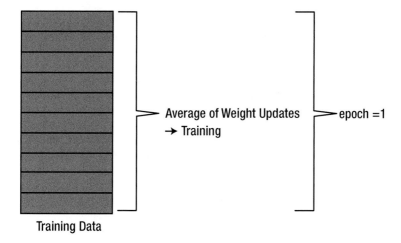

Training Data

Figure 2-18. *The number of training cycles of the neural network equals an epoch*

In contrast, in the mini batch, the number of training processes for one epoch varies depending on the number of data points in each batch. When we have N training data points in total, the number of training processes per epoch is greater than one, which corresponds to the batch method, and smaller than N, which corresponds to the SGD.

Example: Delta Rule

You are now ready to implement the delta rule as a code. Consider a neural network that consists of three input nodes and one output node, as shown in Figure 2-19. The sigmoid function is used for the activation function of the output node.

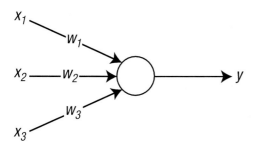

Figure 2-19. *Neural network that consists of three input nodes and one output node*

We have four training data points, as shown in the following table. As they are used for supervised learning, each data point consists of an input-correct output pair. The last bold number of each dataset is the correct output.

{ 0, 0, 1, **0** }
{ 0, 1, 1, **0** }
{ 1, 0, 1, **1** }
{ 1, 1, 1, **1** }

Let's train the neural network with this data. The delta rule for the sigmoid function, which is given by Equation 2.5, is the learning rule. Equation 2.5 can be rearranged as a step-by-step process, as follows:

$$\delta_i = \varphi(v_i)(1-\varphi(v_i))e_i$$
$$\Delta w_{ij} = \alpha\,\delta_i\,x_j \qquad\qquad \text{(Equation 2.7)}$$
$$w_{ij} \leftarrow w_{ij} + \Delta w_{ij}$$

We will implement the delta rule using the SGD and batch methods for the example neural network. As it is single-layered and contains simple training data, the code is not complicated. Once you follow the code, you will clearly see the difference between the SGD code and the batch code. As previously addressed, the SGD trains every data point immediately and does not require addition or averages of the weight updates. Therefore, the code for the SGD is simpler than that of the batch.

Implementation of the SGD Method

The function DeltaSGD is the SGD method of the delta rule given by Equation 2.7. It takes the weights and training data of the neural network and returns the newly trained weights.

```
W = DeltaSGD(W, X, D)
```

where W is the argument that carries the weights. X and D carry the inputs and correct outputs of the training data, respectively. The training data is divided into two variables for convenience. The following listing shows the DeltaSGD.m file, which implements the DeltaSGD function.

```
function W = DeltaSGD(W, X, D)
  alpha = 0.9;

  N = 4;
  for k = 1:N
    x = X(k, :)';
    d = D(k);

    v = W*x;
    y = Sigmoid(v);

    e     = d - y;
    delta = y*(1-y)*e;

    dW = alpha*delta*x;       % delta rule

    W(1) = W(1) + dW(1);
    W(2) = W(2) + dW(2);
    W(3) = W(3) + dW(3);
  end
end
```

The code proceeds as follows: Take one of the data points and calculate the output, y. Calculate the difference between this output and the correct output, d. Calculate the weight update, dW, according to the delta rule. Using this weight update, adjust the weight of neural network. Repeat the process for the number of the training data points, N. This way, the function DeltaSGD trains the neural network for every epoch.

The function Sigmoid that DeltaSGD calls is listed next. This outlines the pure definition of the sigmoid function and is implemented in the Sigmoid.m file. As it is a very simple code, we skip further discussion of it.

```
function y = Sigmoid(x)
  y = 1 / (1 + exp(-x));
end
```

The following listing shows the TestDeltaSGD.m file, which tests the DeltaSGD function. This program calls the function DeltaSGD, trains it 10,000 times, and displays the output from the trained neural network with the input of all the training data. We can see how well the neural network was trained by comparing the output with the correct output.

```
clear all

X = [ 0 0 1;
      0 1 1;
      1 0 1;
      1 1 1;
    ];

D = [ 0
      0
      1
      1
    ];

W = 2*rand(1, 3) - 1;

for epoch = 1:10000              % train
   W = DeltaSGD(W, X, D);
end

N = 4;                          % inference
for k = 1:N
   x = X(k, :)';
   v = W*x;
   y = Sigmoid(v)
end
```

This code initializes the weights with random real numbers between -1 and 1. Executing this code produces the following values. These output values are very close to the correct outputs in D. Therefore, we can conclude that the neural network has been properly trained.

$$\begin{bmatrix} 0.0102 \\ 0.0083 \\ 0.9932 \\ 0.9917 \end{bmatrix} \quad \Leftrightarrow \quad D = \begin{bmatrix} 0 \\ 0 \\ 1 \\ 1 \end{bmatrix}$$

Every example code in this book consists of the implementation of the algorithm and the test program in separate files. This is because putting them together often makes the code more complicated and hampers efficient analysis of the algorithm. The file name of the test program starts with Test and is followed by the name on the algorithm file. The algorithm file is named after the function name, in compliance with the naming convention of MATLAB. For example, the implementation file of the DeltaSGD function is named DeltaSGD.m.

Algorithm implementation	example/ DeltaSGD.m
Test program	example/ TestDeltaSGD.m

Implementation of the Batch Method

The function DeltaBatch implements the delta rule of Equation 2.7 using the batch method. It takes the weights and training data of the neural network and returns trained weights.

```
W = DeltaBatch(W, X, D)
```

In this function definition, the variables carry the same meaning as those in the function DeltaSGD; W is the weight of the neural network, X and D are the input and correct output of the training data, respectively. The following listing shows the DeltaBatch.m file, which implements the function DeltaBatch.

```
function W = DeltaBatch(W, X, D)
  alpha = 0.9;

  dWsum = zeros(3, 1);

  N = 4;
  for k = 1:N
    x = X(k, :)';
    d = D(k);

    v = W*x;
    y = Sigmoid(v);

    e     = d - y;
    delta = y*(1-y)*e;

    dW = alpha*delta*x;

    dWsum = dWsum + dW;
  end
  dWavg = dWsum / N;
```

```
    W(1) = W(1) + dWavg(1);
    W(2) = W(2) + dWavg(2);
    W(3) = W(3) + dWavg(3);
end
```

This code does not immediately train the neural network with the weight update, dW, of the individual training data points. It adds the individual weight updates of the entire training data to dWsum and adjusts the weight just once using the average, dWavg. This is the fundamental difference that separates this method from the SGD method. The averaging feature of the batch method allows the training to be less sensitive to the training data.

Recall that Equation 2.6 yields the weight update. It will be much easier to understand this equation when you look into it using the previous code. Equation 2.6 is shown here again, for your convenience.

$$\Delta w_{ij} = \frac{1}{N} \sum_{k=1}^{N} \Delta w_{ij}(k)$$

where $\Delta w_{ij}(k)$ is the weight update for the k -th training data point.

The following program listing shows the TestDeltaBatch.m file that tests the function DeltaBatch. This program calls in the function DeltaBatch and trains the neural network 40,000 times. All the training data is fed into the trained neural network, and the output is displayed. Check the output and correct output from the training data to verify the adequacy of the training.

```
clear all

X = [ 0 0 1;
      0 1 1;
      1 0 1;
      1 1 1;
    ];

D = [ 0
      0
      1
      1
    ];

W = 2*rand(1, 3) - 1;

for epoch = 1:40000
    W = DeltaBatch(W, X, D);
end
```

```
N = 4;
for k = 1:N
    x = X(k, :)';
    v = W*x;
    y = Sigmoid(v)
end
```

Next, execute this code, and you will see the following values on your screen. The output is very similar to the correct output, D. This verifies that the neural network has been properly trained.

$$
\begin{bmatrix} 0.0102 \\ 0.0083 \\ 0.9932 \\ 0.9917 \end{bmatrix} \Leftrightarrow D = \begin{bmatrix} 0 \\ 0 \\ 1 \\ 1 \end{bmatrix}
$$

As this test program is almost identical to the TestDeltaSGD.m file, we will skip the detailed explanation. An interesting point about this method is that it trained the neural network 40,000 times. Recall that the SGD method performed only 10,000 trainings. This indicates that the batch method requires more time to train the neural network to yield a similar level of accuracy of that of the SGD method. In other words, the batch method learns slowly.

Comparison of the SGD and the Batch

In this section, we practically investigate the learning speeds of the SGD and the batch. The errors of these methods are compared at the end of the training processes for the entire training data. The following program listing shows the SGDvsBatch.m file, which compares the mean error of the two methods. In order to evaluate a fair comparison, the weights of both methods are initialized with the same values.

```
clear all

X = [ 0 0 1;
      0 1 1;
      1 0 1;
      1 1 1;
    ];
```

```
D = [ 0
      0
      1
      1
    ];

E1 = zeros(1000, 1);
E2 = zeros(1000, 1);

W1 = 2*rand(1, 3) - 1;
W2 = W1;

for epoch = 1:1000              % train
  W1 = DeltaSGD(W1, X, D);
  W2 = DeltaBatch(W2, X, D);

  es1 = 0;
  es2 = 0;
  N   = 4;
  for k = 1:N
    x = X(k, :)';
    d = D(k);

    v1  = W1*x;
    y1  = Sigmoid(v1);
    es1 = es1 + (d - y1)^2;

    v2  = W2*x;
    y2  = Sigmoid(v2);
    es2 = es2 + (d - y2)^2;
  end
  E1(epoch) = es1 / N;
  E2(epoch) = es2 / N;
end

plot(E1, 'r')
hold on
plot(E2, 'b:')
xlabel('Epoch')
ylabel('Average of Training error')
legend('SGD', 'Batch')
```

This program trains the neural network 1,000 times for each function, DeltaSGD and DeltaBatch. At each epoch, it inputs the training data into the neural network and calculates the mean square error (E1, E2) of the output. Once the program completes 1,000 trainings, it generates a graph that shows the mean error at each epoch. As Figure 2-20 shows, the SGD yields faster reduction of the learning error than the batch; the SGD learns faster.

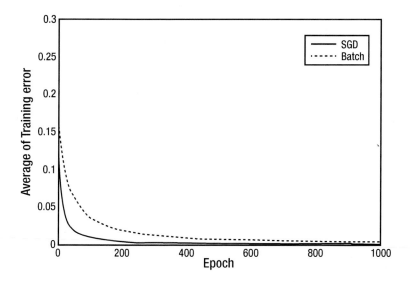

Figure 2-20. *The SGD method learns faster than the batch method*

Limitations of Single-Layer Neural Networks

This section presents the critical reason that the single-layer neural network had to evolve into a multi-layer neural network. We will try to show this through a particular case. Consider the same neural network that was discussed in the previous section. It consists of three input nodes and an output node, and the activation function of the output node is a sigmoid function (Figure 2-21).

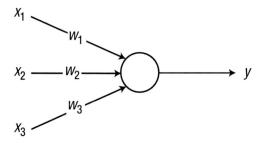

Figure 2-21. *Our same neural network*

Assume that we have four training data points, as shown here. It is different from that of the "Example: Delta Rule" section in that the second and fourth correct outputs are switched while the inputs remain the same. Well, the difference is barely noticeable. It shouldn't cause any trouble, right?

{ 0, 0, 1, **0** }
{ 0, 1, 1, **1** }
{ 1, 0, 1, **1** }
{ 1, 1, 1, **0** }

We will now train it with the delta rule using the SGD. As we are considering the same neural network, we can train it using the function DeltaSGD from the "Example: Delta Rule" section. We have to just change its name to DeltaXOR. The following program listing shows the TestDeltaXOR.m file, which tests the DeltaXOR function. This program is identical to the TestDeltaSGD.m file from the "Example: Delta Rule" section, except that it has different values for D, and it calls the DeltaXOR function instead of DeltaSGD.

```
clear all

X = [ 0 0 1;
      0 1 1;
      1 0 1;
      1 1 1;
    ];

D = [ 0
      1
      1
      0
    ];

W = 2*rand(1, 3) - 1;

for epoch = 1:40000              % train
  W = DeltaXOR(W, X, D);
end

N = 4;                          % inference
for k = 1:N
  x = X(k, :)';
  v = W*x;
  y = Sigmoid(v)
end
```

When we run the code, the screen will show the following values, which consist of the output from the trained neural network corresponding to the training data. We can compare them with the correct outputs given by D.

$$\begin{bmatrix} 0.5297 \\ 0.5000 \\ 0.4703 \\ 0.4409 \end{bmatrix} \quad \Leftrightarrow \quad D = \begin{bmatrix} 0 \\ 1 \\ 1 \\ 0 \end{bmatrix}$$

What happened? We got two totally different sets. Training the neural network for a longer period does not make a difference. The only difference from the code from the "Example: Delta Rule" section is the correct output variable, D. What actually happened?

Illustrating the training data can help elucidate this problem. Let's interpret the three values of the input data as the X, Y, and Z coordinates, respectively. As the third value, i.e. the Z coordinate, is fixed as 1, the training data can be visualized on a plane as shown in Figure 2-22.

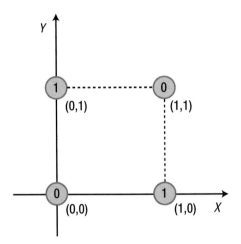

Figure 2-22. *Interpreting the three values of the input data as the X, Y, and Z coordinates*

The values 0 and 1 in the circles are the correct outputs assigned to each point. One thing to notice from this figure is that we cannot divide the regions of 0 and 1 with a straight line. However, we may divide it with a complicated curve, as shown in Figure 2-23. This type of problem is said to be linearly inseparable.

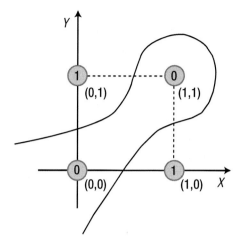

Figure 2-23. *We can only separate the regions of 0 and 1 with a complicated curve*

In the same process, the training data from the "Example: Delta Rule" section on the X-Y plane appears in Figure 2-24.

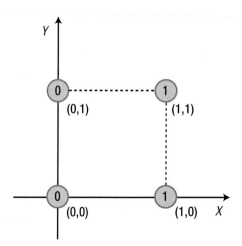

Figure 2-24. *The delta rule training data*

In this case, a straight border line that divides the regions of 0 and 1 can be found easily. This is a linearly separable problem (Figure 2-25).

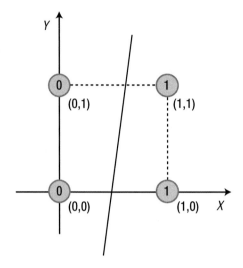

Figure 2-25. *This data presents a linearly separable problem*

To put it simply, the single-layer neural network can only solve linearly separable problems. This is because the single-layer neural network is a model that linearly divides the input data space. In order to overcome this limitation of the single-layer neural network, we need more layers in the network. This need has led to the appearance of the multi-layer neural network, which can achieve what the single-layer neural network cannot. As this is rather mathematical; it is fine to skip this portion if you are not familiar with it. Just keep in mind that the single-layer neural network is applicable for specific problem types. The multi-layer neural network has no such limitations.

Summary

This chapter covered the following concepts:

- The neural network is a network of nodes, which imitate the neurons of the brain. The nodes calculate the weighted sum of the input signals and output the result of the activation function with the weighted sum.

- Most neural networks are constructed with the layered nodes. For the layered neural network, the signal enters through the input layer, passes through the hidden layer, and exits through the output layer.

- In practice, the linear functions cannot be used as the activation functions in the hidden layer. This is because the linear function negates the effects of the hidden layer. However, in some problems such as regression, the output layer nodes may employ linear functions.

- For the neural network, supervised learning implements the process to adjust the weights and to reduce the discrepancies between the correct output and output of the neural network (Figure 2-26).

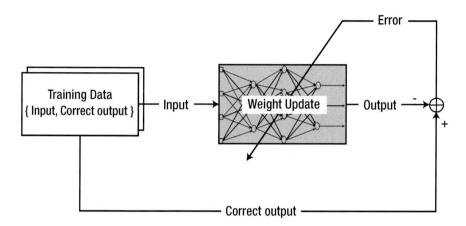

Figure 2-26. *Review of supervised learning*

- The method used to adjust the weight according to the training data is called the learning rule.

- There are three major types of error calculations: the stochastic gradient descent, batch, and mini batch.

- The delta rule is the representative learning rule of the neural network. Its formula varies depending on the activation function.

$$\delta_i = \varphi'(v_i)e_i$$
$$w_{ij} \leftarrow w_{ij} + \alpha \delta_i x_j$$

- The delta rule is an iterative method that gradually reaches the solution. Therefore, the network should be repeatedly trained with the training data until the error is reduced to the satisfactory level.

- The single-layer neural network is applicable only to specific types of problems. Therefore, the single-layer neural network has very limited usages. The multi-layer neural network has been developed to overcome the essential limitations of the single-layer neural network.

CHAPTER 3

■ ■ ■

Training of Multi-Layer Neural Network

In an effort to overcome the practical limitations of the single-layer, the neural network evolved into a multi-layer architecture. However, it has taken approximately 30 years to just add on the hidden layer to the single-layer neural network. It's not easy to understand why this took so long, but the problem involved the learning rule. As the training process is the only method for the neural network to store information, untrainable neural networks are useless. A proper learning rule for the multi-layer neural network took quite some time to develop.

The previously introduced delta rule is ineffective for training of the multi-layer neural network. This is because the error, the essential element for applying the delta rule for training, is not defined in the hidden layers. The error of the output node is defined as the difference between the correct output and the output of the neural network. However, the training data does not provide correct outputs for the hidden layer nodes, and hence the error cannot be calculated using the same approach for the output nodes. Then, what? Isn't the real problem how to define the error at the hidden nodes? You got it. You just formulated the back-propagation algorithm, the representative learning rule of the multi-layer neural network.

In 1986, the introduction of the back-propagation algorithm finally solved the training problem of the multi-layer neural network.[1] The significance of the back-propagation algorithm was that it provided a systematic method to determine the error of the hidden nodes. Once the hidden layer errors are determined, the delta rule is applied to adjust the weights. See Figure 3-1.

---

[1]"Learning representations by back-propagating errors," David E. Rumelhart, Geoffrey E. Hinton, Ronald J. Williams, *Nature*, October 1986.

© Phil Kim 2017
P. Kim, *MATLAB Deep Learning*, DOI 10.1007/978-1-4842-2845-6_3

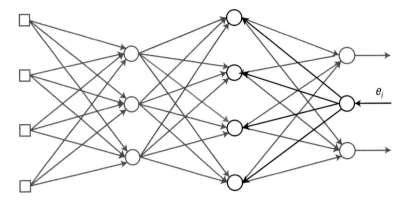

Figure 3-1. *Illustration of back-propagation*

The input data of the neural network travels through the input layer, hidden layer, and output layer. In contrast, in the back-propagation algorithm, the output error starts from the output layer and moves backward until it reaches the right next hidden layer to the input layer. This process is called back-propagation, as it resembles an output error propagating backward. Even in back-propagation, the signal still flows through the connecting lines and the weights are multiplied. The only difference is that the input and output signals flow in opposite directions.

Back-Propagation Algorithm

This section explains the back-propagation algorithm using an example of the simple multi-layer neural network. Consider a neural network that consists of two nodes for both the input and output and a hidden layer, which has two nodes as well. We will omit the bias for convenience. The example neural network is shown in Figure 3-2, where the superscript describes the layer indicator.

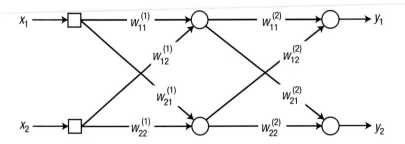

Figure 3-2. *Neural network that consists of two nodes for the input and output and a hidden layer, which has two nodes*

In order to obtain the output error, we first need the neural network's output from the input data. Let's try. As the example network has a single hidden layer, we need two input data manipulations before the output calculation is processed. First, the weighted sum of the hidden node is calculated as:

$$\begin{bmatrix} v_1^{(1)} \\ v_2^{(1)} \end{bmatrix} = \begin{bmatrix} w_{11}^{(1)} & w_{12}^{(1)} \\ w_{21}^{(1)} & w_{22}^{(1)} \end{bmatrix} \begin{bmatrix} x_1 \\ x_2 \end{bmatrix}$$

(Equation 3.1)

$$\triangleq W_1 x$$

When we put this weighted sum, Equation 3.1, into the activation function, we obtain the output from the hidden nodes.

$$\begin{bmatrix} y_1^{(1)} \\ y_2^{(1)} \end{bmatrix} = \begin{bmatrix} \varphi(v_1^{(1)}) \\ \varphi(v_2^{(1)}) \end{bmatrix}$$

where $y_1^{(1)}$ and $y_2^{(1)}$ are outputs from the corresponding hidden nodes. In a similar manner, the weighted sum of the output nodes is calculated as:

$$\begin{bmatrix} v_1 \\ v_2 \end{bmatrix} = \begin{bmatrix} w_{11}^{(2)} & w_{12}^{(2)} \\ w_{21}^{(2)} & w_{22}^{(2)} \end{bmatrix} \begin{bmatrix} y_1^{(1)} \\ y_2^{(1)} \end{bmatrix}$$

(Equation 3.2)

$$\triangleq W_2 y^{(1)}$$

As we put this weighted sum into the activation function, the neural network yields the output.

$$\begin{bmatrix} y_1 \\ y_2 \end{bmatrix} = \begin{bmatrix} \varphi(v_1) \\ \varphi(v_2) \end{bmatrix}$$

Now, we will train the neural network using the back-propagation algorithm. The first thing to calculate is delta, δ, of each node. You may ask, "Is this delta the one from the delta rule?" It is! In order to avoid confusion, the diagram in Figure 3-3 has been redrawn with the unnecessary connections dimmed out.

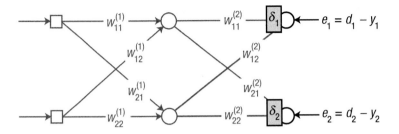

Figure 3-3. *Train the neural network using the back-propagation algorithm*

In the back-propagation algorithm, the delta of the output node is defined identically to the delta rule of the "Generalized Delta Rule" section in Chapter 2, as follows:

$$e_1 = d_1 - y_1$$
$$\delta_1 = \varphi'(v_1)e_1$$

$$e_2 = d_2 - y_2$$
$$\delta_2 = \varphi'(v_2)e_2 \qquad \text{(Equation 3.3)}$$

where $\varphi'(\cdot)$ is the derivative of the activation function of the output node, y_i is the output from the output node, d_i is the correct output from the training data, and v_i is the weighted sum of the corresponding node.

Since we have the delta for every output node, let's proceed leftward to the hidden nodes and calculate the delta (Figure 3-4). Again, unnecessary connections are dimmed out for convenience.

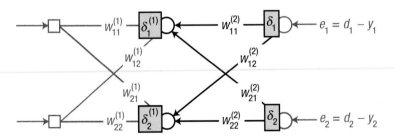

Figure 3-4. *Proceed leftward to the hidden nodes and calculate the delta*

As addressed at the beginning of the chapter, the issue of the hidden node is how to define the error. In the back-propagation algorithm, the error of the node is defined as the weighted sum of the back-propagated deltas from the layer on the immediate right (in this case, the output layer). Once the error is obtained, the calculation of the delta from the node is the same as that of Equation 3.3. This process can be expressed as follows:

$$e_1^{(1)} = w_{11}^{(2)}\delta_1 + w_{21}^{(2)}\delta_2$$

$$\delta_1^{(1)} = \varphi'\left(v_1^{(1)}\right)e_1^{(1)}$$

$$e_2^{(1)} = w_{12}^{(2)}\delta_1 + w_{22}^{(2)}\delta_2$$

$$\delta_2^{(1)} = \varphi'\left(v_2^{(1)}\right)e_2^{(1)} \qquad \text{(Equation 3.4)}$$

where $v_1^{(1)}$ and $v_2^{(1)}$ are the weight sums of the forward signals at the respective nodes. It is noticeable from this equation that the forward and backward processes are identically applied to the hidden nodes as well as the output nodes. This implies that the output and hidden nodes experience the same backward process. The only difference is the error calculation (Figure 3-5).

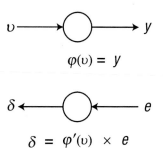

Figure 3-5. *The error calculation is the only difference*

In summary, the error of the hidden node is calculated as the backward weighted sum of the delta, and the delta of the node is the product of the error and the derivative of the activation function. This process begins at the output layer and repeats for all hidden layers. This pretty much explains what the back-propagation algorithm is about.

The two error calculation formulas of Equation 3.4 are combined in a matrix equation as follows:

$$\begin{bmatrix} e_1^{(1)} \\ e_2^{(1)} \end{bmatrix} = \begin{bmatrix} w_{11}^{(2)} & w_{21}^{(2)} \\ w_{12}^{(2)} & w_{22}^{(2)} \end{bmatrix} \begin{bmatrix} \delta_1 \\ \delta_2 \end{bmatrix} \qquad \text{(Equation 3.5)}$$

Compare this equation with the neural network output of Equation 3.2. The matrix of Equation 3.5 is the result of *transpose* of the weight matrix, W, of Equation 3.2.[2] Therefore, Equation 3.5 can be rewritten as:

$$\begin{bmatrix} e_1^{(1)} \\ e_2^{(1)} \end{bmatrix} = W_2^T \begin{bmatrix} \delta_1 \\ \delta_2 \end{bmatrix} \qquad \text{(Equation 3.6)}$$

This equation indicates that we can obtain the error as the product of the transposed weight matrix and delta vector. This very useful attribute allows an easier implementation of the algorithm.

If we have additional hidden layers, we will just repeat the same backward process for each hidden layer and calculate all the deltas. Once all the deltas have been calculated, we will be ready to train the neural network. Just use the following equation to adjust the weights of the respective layers.

$$\Delta w_{ij} = \alpha \delta_i x_j \qquad \text{(Equation 3.7)}$$
$$w_{ij} \leftarrow w_{ij} + \Delta w_{ij}$$

where x_j is the input signal for the corresponding weight. For convenience, we omit the layer indicator from this equation. What do you see now? Isn't this equation the same as that of the delta rule of the previous section? Yes, they are the same. The mere difference is the deltas of the hidden nodes, which are obtained from the backward calculation using the output error of the neural network.

We will proceed a bit further and derive the equation to adjust the weight using Equation 3.7. Consider the weight $w_{21}^{(2)}$ for example.

The weight $w_{21}^{(2)}$ of Figure 3-6 can be adjusted using Equation 3.7 as:

---

[2]When two matrices have rows and columns switched, they are transpose matrices to each other.

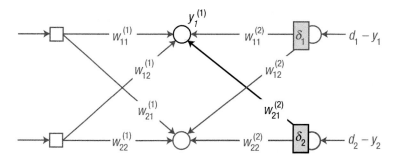

Figure 3-6. *Derive the equation to adjust the weight*

$$w_{21}^{(2)} \leftarrow w_{21}^{(2)} + \alpha\delta_2 y_1^{(1)}$$

where $y_1^{(1)}$ is the output of the first hidden node. Here is another example. The weight $w_{11}^{(1)}$ of Figure 3-7 is adjusted using Equation 3.7 as:

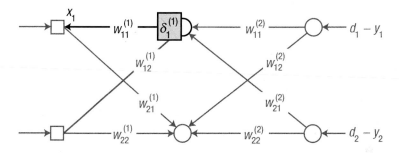

Figure 3-7. *Derive the equation to adjust the weight, again*

$$w_{11}^{(1)} \leftarrow w_{11}^{(1)} + \alpha\delta_1^{(1)} x_1$$

where x_1 is the output of the first input node, i.e., the first input of the neural network.

Let's organize the process to train the neural network using the back-propagation algorithm.

1.   Initialize the weights with adequate values.

2.   Enter the input from the training data { input, correct output } and obtain the neural network's output.
     Calculate the error of the output to the correct output and the delta, δ, of the output nodes.

$$e = d - y$$
$$\delta = \varphi'(v)e$$

3.   Propagate the output node delta, δ, backward, and calculate the deltas of the immediate next (left) nodes.

$$e^{(k)} = W^T \delta$$
$$\delta^{(k)} = \varphi'\left(v^{(k)}\right)e^{(k)}$$

4.   Repeat Step 3 until it reaches the hidden layer that is on the immediate right of the input layer.

5.   Adjust the weights according to the following learning rule.

$$\Delta w_{ij} = \alpha \delta_i x_j$$
$$w_{ij} \leftarrow w_{ij} + \Delta w_{ij}$$

6.   Repeat Steps 2-5 for every training data point.

7.   Repeat Steps 2-6 until the neural network is properly trained.

Other than Steps 3 and 4, in which the output delta propagates backward to obtain the hidden node delta, this process is basically the same as that of the delta rule, which was previously discussed. Although this example has only one hidden layer, the back-propagation algorithm is applicable for training many hidden layers. Just repeat Step 3 of the previous algorithm for each hidden layer.

Example: Back-Propagation

In this section, we implement the back-propagation algorithm. The training data contains four elements as shown in the following table. Of course, as this is about supervised learning, the data includes input and correct output pairs. The bolded rightmost number of the data is the correct output. As you may have noticed, this data is the same one that we used in Chapter 2 for the training of the single-layer neural network; the one that the single-layer neural network had failed to learn.

{ 0, 0, 1, **0** }
{ 0, 1, 1, **1** }
{ 1, 0, 1, **1** }
{ 1, 1, 1, **0** }

Ignoring the third value, the Z-axis, of the input, this dataset actually provides the XOR logic operations. Therefore, if we train the neural network with this dataset, we would get the XOR operation model.

Consider a neural network that consists of three input nodes and a single output node, as shown in Figure 3-8. It has one hidden layer of four nodes. The sigmoid function is used as the activation function for the hidden nodes and the output node.

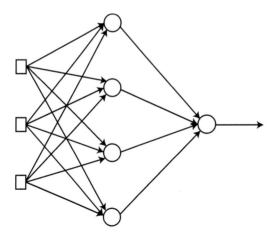

Figure 3-8. *Neural network that consists of three input nodes and a single output node*

This section employs SGD for the implementation of the back-propagation algorithm. Of course, the batch method will work as well. What we have to do is use the average of the weight updates, as shown in the example in the "Example: Delta Rule" section of Chapter 2. Since the primary objective of this section is to understand the back-propagation algorithm, we will stick to a simpler and more intuitive method: the SGD.

XOR Problem

The function BackpropXOR, which implements the back-propagation algorithm using the SGD method, takes the network's weights and training data and returns the adjusted weights.

```
[W1 W2] = BackpropXOR(W1, W2, X, D)
```

where W1 and W2 carries the weight matrix of the respective layer. W1 is the weight matrix between the input layer and hidden layer and W2 is the weight matrix between the hidden layer and output layer. X and D are the input and correct output of the training data, respectively. The following listing shows the BackpropXOR.m file, which implements the BackpropXOR function.

```
function [W1, W2] = BackpropXOR(W1, W2, X, D)
  alpha = 0.9;

  N = 4;
  for k = 1:N
    x = X(k, :)';
    d = D(k);

    v1 = W1*x;
    y1 = Sigmoid(v1);
    v  = W2*y1;
    y  = Sigmoid(v);

    e     = d - y;
    delta = y.*(1-y).*e;

    e1     = W2'*delta;
    delta1 = y1.*(1-y1).*e1;

    dW1 = alpha*delta1*x';
    W1  = W1 + dW1;
```

```
    dW2 = alpha*delta*y1';
    W2  = W2 + dW2;
  end
end
```

The code takes point from the training dataset, calculates the weight update, dW, using the delta rule, and adjusts the weights. So far, the process is almost identical to that of the example code of Chapter 2. The slight differences are the two calls of the function Sigmoid for the output calculation and the addition of the delta (delta1) calculation using the back-propagation of the output delta as follows:

```
e1     = W2'*delta;
delta1 = y1.*(1-y1).*e1;
```

where the calculation of the error, e1, is the implementation of Equation 3.6. As this involves the back-propagation of the delta, we use the transpose matrix, W2'. The delta (delta1) calculation has an element-wise product operator, .*, because the variables are vectors. The element-wise operator of MATLAB has a dot (period) in front of the normal operator and performs an operation on each element of the vector. This operator enables simultaneous calculations of deltas from many nodes.

The function Sigmoid, which the BackpropXOR code calls, also replaced the division with the element-wise division ./ to account for the vector.

```
function y = Sigmoid(x)
  y = 1 ./ (1 + exp(-x));
end
```

The modified Sigmoid function can operate using vectors as shown by the following example:

```
Sigmoid([-1 0 1])    ➜    [0.2689   0.5000   0.7311]
```

The program listing that follows shows the TestBackpropXOR.m file, which tests the function BackpropXOR. This program calls in the BackpropXOR function and trains the neural network 10,000 times. The input is given to the trained network, and its output is shown on the screen. The training performance can be verified as we compare the output to the correct outputs of the training data. Further details are omitted, as the program is almost identical to that of Chapter 2.

```
clear all

X = [ 0 0 1;
      0 1 1;
      1 0 1;
      1 1 1;
    ];

D = [ 0
      1
      1
      0
    ];

W1 = 2*rand(4, 3) - 1;
W2 = 2*rand(1, 4) - 1;

for epoch = 1:10000            % train
  [W1 W2] = BackpropXOR(W1, W2, X, D);
end

N = 4;                         % inference
for k = 1:N
  x  = X(k, :)';
  v1 = W1*x;
  y1 = Sigmoid(v1);
  v  = W2*y1;
  y  = Sigmoid(v)
end
```

Execute the code, and find the following values on the screen. These values are very close to the correct output, D, indicating that the neural network has been properly trained. Now we have confirmed that the multi-layer neural network solves the XOR problem, which the single-layer network had failed to model properly.

$$\begin{bmatrix} 0.0060 \\ 0.9888 \\ 0.9891 \\ 0.0134 \end{bmatrix} \Leftrightarrow D = \begin{bmatrix} 0 \\ 1 \\ 1 \\ 0 \end{bmatrix}$$

Momentum

This section explores the variations of the weight adjustment. So far, the weight adjustment has relied on the simplest forms of Equations 2.7 and 3.7. However, there are various weight adjustment forms available.[3] The benefits of using the advanced weight adjustment formulas include higher stability and faster speeds in the training process of the neural network. These characteristics are especially favorable for Deep Learning as it is hard to train. This section only covers the formulas that contain momentum, which have been used for a long time. If necessary, you may want to study this further with the link shown in the footnote.

The momentum, m, is a term that is added to the delta rule for adjusting the weight. The use of the momentum term drives the weight adjustment to a certain direction to some extent, rather than producing an immediate change. It acts similarly to physical momentum, which impedes the reaction of the body to the external forces.

$$\Delta w = \alpha \delta x$$
$$m = \Delta w + \beta m^{-}$$
$$w = w + m \qquad \text{(Equation 3.8)}$$
$$m^{-} = m$$

where m^{-} is the previous momentum and β is a positive constant that is less than 1. Let's briefly see why we modify the weight adjustment formula in this manner. The following steps show how the momentum changes over time:

$$m(0) = 0$$
$$m(1) = \Delta w(1) + \beta m(0) = \Delta w(1)$$
$$m(2) = \Delta w(2) + \beta m(1) = \Delta w(2) + \beta \Delta w(1)$$
$$m(3) = \Delta w(3) + \beta m(2) = \Delta w(3) + \beta \{\Delta w(2) + \beta \Delta w(1)\}$$
$$= \Delta w(3) + \beta \Delta w(2) + \beta^{2} \Delta w(1)$$
$$\vdots$$

It is noticeable from these steps that the previous weight update, i.e. $\Delta w(1)$, $\Delta w(2)$, $\Delta w(3)$, etc., is added to each momentum over the process. Since β is less than 1, the older weight update exerts a lesser influence on the momentum. Although the influence diminishes over time, the old weight updates remain

---

[3]sebastianruder.com/optimizing-gradient-descent

in the momentum. Therefore, the weight is not solely affected by a particular weight update value. Therefore, the learning stability improves. In addition, the momentum grows more and more with weight updates. As a result, the weight update becomes greater and greater as well. Therefore, the learning rate increases.

The following listing shows the BackpropMmt.m file, which implements the back-propagation algorithm with the momentum. The BackpropMmt function operates in the same manner as that of the previous example; it takes the weights and training data and returns the adjusted weights. This listing employs the same variables as defined in the BackpropXOR function.

```
[W1 W2] = BackpropMmt(W1, W2, X, D)
function [W1, W2] = BackpropMmt(W1, W2, X, D)
  alpha = 0.9;
  beta  = 0.9;

  mmt1 = zeros(size(W1));
  mmt2 = zeros(size(W2));

  N = 4;
  for k = 1:N
    x = X(k, :)';
    d = D(k);

    v1 = W1*x;
    y1 = Sigmoid(v1);
    v  = W2*y1;
    y  = Sigmoid(v);

    e     = d - y;
    delta = y.*(1-y).*e;

    e1      = W2'*delta;
    delta1 = y1.*(1-y1).*e1;

    dW1  = alpha*delta1*x';
    mmt1 = dW1 + beta*mmt1;
    W1   = W1 + mmt1;

    dW2  = alpha*delta*y1';
    mmt2 = dW2 + beta*mmt2;
    W2   = W2 + mmt2;
  end
end
```

The code initializes the momentums, mmt1 and mmt2, as zeros when it starts the learning process. The weight adjustment formula is modified to reflect the momentum as:

```
dW1   = alpha*delta1*x';
mmt1 = dW1 + beta*mmt1;
W1    = W1 + mmt1;
```

The following program listing shows the TestBackpropMmt.m file, which tests the function BackpropMmt. This program calls the BackpropMmt function and trains the neural network 10,000 times. The training data is fed to the neural network and the output is shown on the screen. The performance of the training is verified by comparing the output to the correct output of the training data. As this code is almost identical to that of the previous example, further explanation is omitted.

```
clear all

X = [ 0 0 1;
      0 1 1;
      1 0 1;
      1 1 1;
    ];

D = [ 0
      1
      1
      0
    ];

W1 = 2*rand(4, 3) - 1;
W2 = 2*rand(1, 4) - 1;

for epoch = 1:10000          % train
  [W1 W2] = BackpropMmt(W1, W2, X, D);
end

N = 4;                       % inference
for k = 1:N
  x  = X(k, :)';
  v1 = W1*x;
  y1 = Sigmoid(v1);
  v  = W2*y1;
  y  = Sigmoid(v)
end
```

Cost Function and Learning Rule

This section briefly explains what the *cost function*[4] is and how it affects the learning rule of the neural network. The cost function is a rather mathematical concept that is associated with the optimization theory. You don't have to know it. However, it is good to know if you want to better understand the learning rule of the neural network. It is not a difficult concept to follow.

The cost function is related to supervised learning of the neural network. Chapter 2 addressed that supervised learning of the neural network is a process of adjusting the weights to reduce the error of the training data. In this context, the measure of the neural network's error is the cost function. The greater the error of the neural network, the higher the value of the cost function is. There are two primary types of cost functions for the neural network's supervised learning.

$$J = \sum_{i=1}^{M} \frac{1}{2} (d_i - y_i)^2 \qquad \text{(Equation 3.9)}$$

$$J = \sum_{i=1}^{M} \{ -d_i \ln(y_i) - (1-d_i) \ln(1-y_i) \} \qquad \text{(Equation 3.10)}$$

where y_i is the output from the output node, d_i is the correct output from the training data, and M is the number of output nodes.

First, consider the sum of squared error shown in Equation 3.9. This cost function is the square of the difference between the neural network's output, y, and the correct output, d. If the output and correct output are the same, the error becomes zero. In contrast, a greater difference between the two values leads to a larger error. This is illustrated in Figure 3-9.

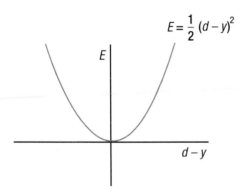

Figure 3-9. *The greater the difference between the output and the correct output, the larger the error*

---

[4]It is also called the loss function and objective function.

It is clearly noticeable that the cost function value is proportional to the error. This relationship is so intuitive that no further explanation is necessary. Most early studies of the neural network employed this cost function to derive learning rules. Not only was the delta rule of the previous chapter derived from this function, but the back-propagation algorithm was as well. Regression problems still use this cost function.

Now, consider the cost function of Equation 3.10. The following formula, which is inside the curly braces, is called the cross entropy function.

$$E = -d\ln(y) - (1-d)\ln(1-y)$$

It may be difficult to intuitively capture the cross entropy function's relationship to the error. This is because the equation is contracted for simpler expression. Equation 3.10 is the concatenation of the following two equations:

$$E = \begin{cases} -\ln(y) & d = 1 \\ -\ln(1-y) & d = 0 \end{cases}$$

Due to the definition of a logarithm, the output, y, should be within 0 and 1. Therefore, the cross entropy cost function often teams up with sigmoid and softmax activation functions in the neural network.[5] Now we will see how this function is related to the error. Recall that cost functions should be proportional to the output error. What about this one?

Figure 3-10 shows the cross entropy function at $d = 1$.

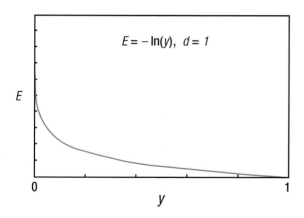

Figure 3-10. The cross entropy function at d = 1

[5]If the other activation function is employed, the definition of the cross entropy function slightly changes as well.

When the output y is 1, i.e., the error ($d - y$) is 0, the cost function value is 0 as well. In contrast, when the output y approaches 0, i.e., the error grows, the cost function value soars. Therefore, this cost function is proportional to the error.

Figure 3-11 shows the cost function at $d = 0$. If the output y is 0, the error is 0, the cost function yields 0. When the output approaches 1, i.e., the error grows, the function value soars. Therefore, this cost function in this case is proportional to the error as well. These cases confirm that the cost function of Equation 3.10 is proportional to the output error of the neural network.

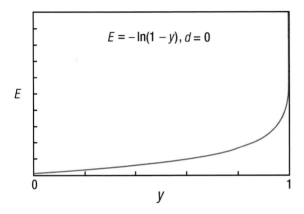

Figure 3-11. *The cross entropy function at d = 0*

The primary difference of the cross entropy function from the quadratic function of Equation 3.9 is its geometric increase. In other words, the cross entropy function is much more sensitive to the error. For this reason, the learning rules derived from the cross entropy function are generally known to yield better performance. It is recommended that you use the cross entropy-driven learning rules except for inevitable cases such as the regression.

We had a long introduction to the cost function because the selection of the cost function affects the learning rule, i.e., the formula of the back-propagation algorithm. Specifically, the calculation of the delta at the output node changes slightly. The following steps detail the procedure in training the neural network with the sigmoid activation function at the output node using the cross entropy-driven back-propagation algorithm.

1. Initialize the neural network's weights with adequate values.

2. Enter the input of the training data { input, correct output } to the neural network and obtain the output. Compare this output to the correct output, calculate the error, and calculate the delta, δ, of the output nodes.

$$e = d - y$$
$$\delta = e$$

3. Propagate the delta of the output node backward and calculate the delta of the subsequent hidden nodes.

$$e^{(k)} = W^T \delta$$
$$\delta^{(k)} = \varphi'\left(v^{(k)}\right)e^{(k)}$$

4. Repeat Step 3 until it reaches the hidden layer that is next to the input layer.

5. Adjust the neural network's weights using the following learning rule:

$$\Delta w_{ij} = \alpha \delta_i x_j$$
$$w_{ij} \leftarrow w_{ij} + \Delta w_{ij}$$

6. Repeat Steps 2-5 for every training data point.

7. Repeat Steps 2-6 until the network has been adequately trained.

Did you notice the difference between this process and that of the "Back-Propagation Algorithm" section? It is the delta, δ, in Step 2. It has been changed as follows:

$$\delta = \varphi'(v)e \quad \rightarrow \quad \delta = e$$

Everything else remains the same. On the outside, the difference seems insignificant. However, it contains the huge topic of the cost function based on the optimization theory. Most of the neural network training approaches of Deep Learning employ the cross entropy-driven learning rules. This is due to their superior learning rate and performance.

Figure 3-12 depicts what this section has explained so far. The key is the fact that the output and hidden layers employ the different formulas of the delta calculation when the learning rule is based on the cross entropy and the sigmoid function.

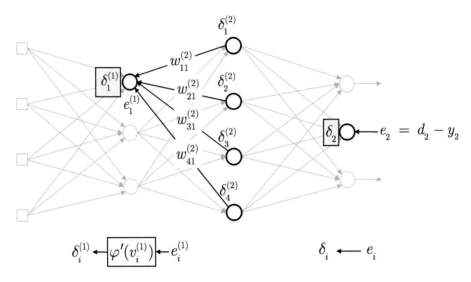

Figure 3-12. *the output and hidden layers employ the different formulas of the delta calculation*

While we are at it, we will address just one more thing about the cost function. You saw in Chapter 1 that overfitting is a challenging problem that every technique of Machine Learning faces. You also saw that one of the primary approaches used to overcome overfitting is making the model as simple as possible using regularization. In a mathematical sense, the essence of regularization is adding the sum of the weights to the cost function, as shown here. Of course, applying the following new cost function leads to a different learning rule formula.

$$J = \frac{1}{2}\sum_{i=1}^{M}(d_i - y_i)^2 + \lambda\frac{1}{2}\|w\|^2$$

$$J = \sum_{i=1}^{M}\{-d_i\ln(y_i) - (1-d_i)\ln(1-y_i)\} + \lambda\frac{1}{2}\|w\|^2$$

where λ is the coefficient that determines how much of the connection weight is reflected on the cost function.

This cost function maintains a large value when one of the output errors and the weight remain large. Therefore, solely making the output error zero will not suffice in reducing the cost function. In order to drop the value of the cost function, both the error and weight should be controlled to be as small as possible. However, if a weight becomes small enough, the associated nodes will be practically disconnected. As a result, unnecessary connections are eliminated, and the neural network becomes simpler. For this reason, overfitting of the neural network can be improved by adding the sum of weights to the cost function, thereby reducing it.

In summary, the learning rule of the neural network's supervised learning is derived from the cost function. The performance of the learning rule and the neural network varies depending on the selection of the cost function. The cross entropy function has been attracting recent attention for the cost function. The regularization process that is used to deal with overfitting is implemented as a variation of the cost function.

Example: Cross Entropy Function

This section revisits the back-propagation example. But this time, the learning rule derived from the cross entropy function is used. Consider the training of the neural network that consists of a hidden layer with four nodes, three input nodes, and a single output node. The sigmoid function is employed for the activation function of the hidden nodes and output node.

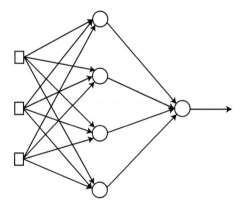

Figure 3-13. *Neural network with a hidden layer with four nodes, three input nodes, and a single output node*

The training data contains the same four elements as shown in the following table. When we ignore the third numbers of the input data, this training dataset presents a XOR logic operation. The bolded rightmost number of each element is the correct output.

<table>
<tr><td>{ 0, 0, 1, 0 }</td></tr>
<tr><td>{ 0, 1, 1, 1 }</td></tr>
<tr><td>{ 1, 0, 1, 1 }</td></tr>
<tr><td>{ 1, 1, 1, 0 }</td></tr>
</table>

Cross Entropy Function

The BackpropCE function trains the XOR data using the cross entropy function. It takes the neural network's weights and training data and returns the adjusted weights.

```
[W1 W2] = BackpropCE(W1, W2, X, D)
```

where W1 and W2 are the weight matrices for the input-hidden layers and hidden-output layers, respectively. In addition, X and D are the input and correct output matrices of the data, respectively. The following listing shows the BackpropCE.m file, which implements the BackpropCE function.

```
function [W1, W2] = BackpropCE(W1, W2, X, D)
  alpha = 0.9;

  N = 4;
  for k = 1:N
    x = X(k, :)';        % x = a column vector
    d = D(k);

    v1 = W1*x;
    y1 = Sigmoid(v1);
    v  = W2*y1;
    y  = Sigmoid(v);

    e     = d - y;
    delta = e;
```

```
    e1      = W2'*delta;
    delta1 = y1.*(1-y1).*e1;

    dW1 = alpha*delta1*x';
    W1 = W1 + dW1;

    dW2 = alpha*delta*y1';
    W2 = W2 + dW2;
  end
end
```

This code pulls out the training data, calculates the weight updates (dW1 and dW2) using the delta rule, and adjusts the neural network's weights using these values. So far, the process is almost identical to that of the previous example. The difference arises when we calculate the delta of the output node as:

```
e       = d - y;
delta = e;
```

Unlike the previous example code, the derivative of the sigmoid function no longer exists. This is because, for the learning rule of the cross entropy function, if the activation function of the output node is the sigmoid, the delta equals the output error. Of course, the hidden nodes follow the same process that is used by the previous back-propagation algorithm.

```
e1      = W2'*delta;
delta1 = y1.*(1-y1).*e1;
```

The following program listing shows the TestBackpropCE.m file, which tests the BackpropCE function. This program calls the BackpropCE function and trains the neural network 10,000 times. The trained neural network yields the output for the training data input, and the result is displayed on the screen. We verify the proper training of the neural network by comparing the output to the correct output. Further explanation is omitted, as the code is almost identical to that from before.

```
clear all

X = [ 0 0 1;
      0 1 1;
      1 0 1;
      1 1 1;
    ];
```

```
D = [ 0
      1
      1
      0
    ];

W1 = 2*rand(4, 3) - 1;
W2 = 2*rand(1, 4) - 1;

for epoch = 1:10000                     % train
   [W1 W2] = BackpropCE(W1, W2, X, D);
end

N = 4;                                  % inference
for k = 1:N
   x  = X(k, :)';
   v1 = W1*x;
   y1 = Sigmoid(v1);
   v  = W2*y1;
   y  = Sigmoid(v)
end
```

Executing this code produces the values shown here. The output is very close to the correct output, D. This proves that the neural network has been trained successfully.

$$\begin{bmatrix} 0.00003 \\ 0.9999 \\ 0.9998 \\ 0.00036 \end{bmatrix} \Leftrightarrow D = \begin{bmatrix} 0 \\ 1 \\ 1 \\ 0 \end{bmatrix}$$

Comparison of Cost Functions

The only difference between the BackpropCE function from the previous section and the BackpropXOR function from the "XOR Problem" section is the calculation of the output node delta. We will examine how this insignificant difference affects the learning performance. The following listing shows the CEvsSSE.m file that compares the mean errors of the two functions. The architecture of this file is almost identical to that of the SGDvsBatch.m file in the "Comparison of the SGD and the Batch" section in Chapter 2.

```
clear all

X = [ 0 0 1;
      0 1 1;
      1 0 1;
      1 1 1;
    ];

D = [ 0
      0
      1
      1
    ];

E1 = zeros(1000, 1);
E2 = zeros(1000, 1);

W11 = 2*rand(4, 3) - 1;      % Cross entropy
W12 = 2*rand(1, 4) - 1;      %
W21 = W11;                   % Sum of squared error
W22 = W12;                   %

for epoch = 1:1000
  [W11 W12] = BackpropCE(W11, W12, X, D);
  [W21 W22] = BackpropXOR(W21, W22, X, D);

  es1 = 0;
  es2 = 0;
  N   = 4;
  for k = 1:N
    x = X(k, :)';
    d = D(k);

    v1  = W11*x;
    y1  = Sigmoid(v1);
    v   = W12*y1;
    y   = Sigmoid(v);
    es1 = es1 + (d - y)^2;

    v1  = W21*x;
    y1  = Sigmoid(v1);
    v   = W22*y1;
```

```
    y   = Sigmoid(v);
    es2 = es2 + (d - y)^2;
  end
  E1(epoch) = es1 / N;
  E2(epoch) = es2 / N;
end

plot(E1, 'r')
hold on
plot(E2, 'b:')
xlabel('Epoch')
ylabel('Average of Training error')
legend('Cross Entropy', 'Sum of Squared Error')
```

This program calls the BackpropCE and the BackpropXOR functions and trains the neural networks 1,000 times each. The squared sum of the output error (es1 and es2) is calculated at every epoch for each neural network, and their average (**E1** and **E2**) is calculated. W11, W12, W21, and W22 are the weight matrices of respective neural networks. Once the 1,000 trainings have been completed, the mean errors are compared over the epoch on the graph. As Figure 3-14 shows, the cross entropy-driven training reduces the training error at a much faster rate. In other words, the cross entropy-driven learning rule yields a faster learning process. This is the reason that most cost functions for Deep Learning employ the cross entropy function.

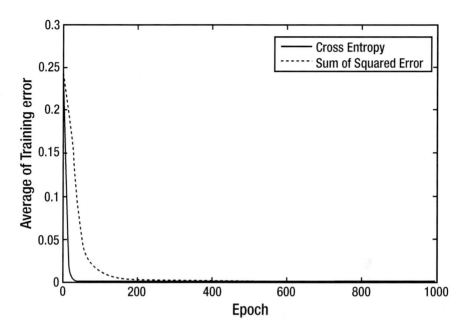

Figure 3-14. *Cross entropy-driven training reduces training error at a much faster rate*

This completes the contents for the back-propagation algorithm. If you had a hard time catching on, don't be discouraged. Actually, understanding the back-propagation algorithms is not a vital factor when studying and developing Deep Learning. As most of the Deep Learning libraries already include the algorithms; we can just use them. Cheer up! Deep Learning is just one chapter away.

Summary

This chapter covered the following concepts:

- The multi-layer neural network cannot be trained using the delta rule; it should be trained using the back-propagation algorithm, which is also employed as the learning rule of Deep Learning.

- The back-propagation algorithm defines the hidden layer error as it propagates the output error backward from the output layer. Once the hidden layer error is obtained, the weights of every layer are adjusted using the delta rule. The importance of the back-propagation algorithm is that it provides a systematic method to define the error of the hidden node.

- The single-layer neural network is applicable only to linearly separable problems, and most practical problems are linearly inseparable.

- The multi-layer neural network is capable of modeling the linearly inseparable problems.

- Many types of weight adjustments are available in the back-propagation algorithm. The development of various weight adjustment approaches is due to the pursuit of a more stable and faster learning of the network. These characteristics are particularly beneficial for hard-to-learn Deep Learning.

- The cost function addresses the output error of the neural network and is proportional to the error. Cross entropy has been widely used in recent applications. In most cases, the cross entropy-driven learning rules are known to yield better performance.

- The learning rule of the neural network varies depending on the cost function and activation function. Specifically, the delta calculation of the output node is changed.

- The regularization, which is one of the approaches used to overcome overfitting, is also implemented as an addition of the weight term to the cost function.

CHAPTER 4

■ ■ ■

Neural Network and Classification

As addressed in Chapter 1, the primary Machine Learning applications that require supervised learning are classification and regression. Classification is used to determine the group the data belongs. Some typical applications of classification are spam mail filtering and character recognition. In contrast, regression infers values from the data. It can be exemplified with the prediction of income for a given age and education level.

Although the neural network is applicable to both classification and regression, it is seldom used for regression. This is not because it yields poor performance, but because most of regression problems can be solved using simpler models. Therefore, we will stick to classification throughout this book.

In the application of the neural network to classification, the output layer is usually formulated differently depending on how many groups the data should be divided into. The selection of the number of nodes and suitable activation functions for the classification of two groups is different when using more groups. Keep in mind that it affects only the output nodes, while the hidden nodes remain intact. Of course, the approaches of this chapter are not only ones available. However, these may be the best to start with, as they have been validated through many studies and cases.

Binary Classification

We will start with the binary classification neural network, which classifies the input data into one of the two groups. This kind of classifier is actually useful for more applications than you may expect. Some typical applications include spam mail filtering (a spam mail or a normal mail) and loan approvals (approve or deny).

P. Kim, *MATLAB Deep Learning*, DOI 10.1007/978-1-4842-2845-6_4

For binary classification, a single output node is sufficient for the neural network. This is because the input data can be classified by the output value, which is either greater than or less than the threshold. For example, if the sigmoid function is employed as the activation function of the output node, the data can be classified by whether the output is greater than 0.5 or not. As the sigmoid function ranges from 0-1, we can divide groups in the middle, as shown in Figure 4-1.

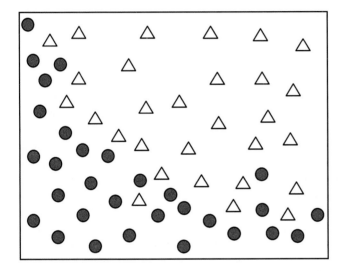

Figure 4-1. Binary classification problem

Consider the binary classification problem shown in Figure 4-1. For the given coordinates (x, y), the model is to determine which group the data belongs. In this case, the training data is given in the format shown in Figure 4-2. The first two numbers indicate the x and y coordinates respectively, and the symbol represents the group in which the data belongs. The data consists of the input and correct output as it is used for supervised learning.

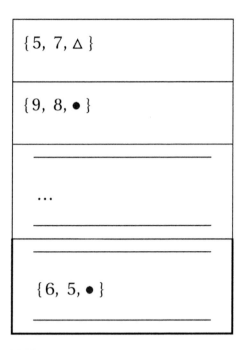

Figure 4-2. *Training data binary classification*

Now, let's construct the neural network. The number of input nodes equals the number of input parameters. As the input of this example consists of two parameters, the network employs two input nodes. We need one output node because this implements the classification of two groups as previously addressed. The sigmoid function is used as the activation function, and the hidden layer has four nodes.[1] Figure 4-3 shows the described neural network.

---

[1]The hidden layer is not our concern. The layer that varies depending on the number of classes is the output layer, not the hidden layer. There is no standard rule for the composition of the hidden layer.

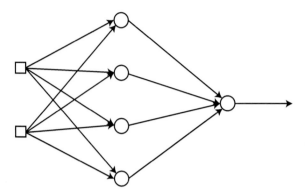

Figure 4-3. *Neural network for the training data*

When we train this network with the given training data, we can get the binary classification that we want. However, there is a problem. The neural network produces numerical outputs that range from 0-1, while we have the symbolic correct outputs given as △ and ●. We cannot calculate the error in this way; we need to switch the symbols to numerical codes. We can assign the maximum and minimum values of the sigmoid function to the two classes as follows:

Class △ → 1

Class ● → 0

The change of the class symbols yields the training data shown in Figure 4-4.

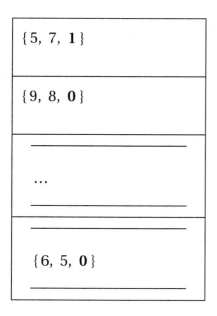

Figure 4-4. *Change the class symbols and the data is classified differently*

The training data shown in Figure 4-4 is what we use to train the neural network. The binary classification neural network usually adopts the cross entropy function of the previous equation for training. You don't have to do so all the time, but it is beneficial for the performance and implementation process. The learning process of the binary classification neural network is summarized in the following steps. Of course, we use the cross entropy function as the cost function and the sigmoid function as the activation function of the hidden and output nodes.

1. The binary classification neural network has one node for the output layer. The sigmoid function is used for the activation function.

2. Switch the class titles of the training data into numbers using the maximum and minimum values of the sigmoid function.

   Class $\triangle$ → 1

   Class ● → 0

3. Initialize the weights of the neural network with adequate values.

4. Enter the input from the training data { input, correct output } into the neural network and obtain the output. Calculate the error between the output and correct output, and determine the delta, δ, of the output nodes.

$$e = d - y$$
$$d = e$$

5. Propagate the output delta backwards and calculate the delta of the subsequent hidden nodes.

$$e^{(k)} = W^T \delta$$
$$\delta^{(k)} = \varphi'\left(v^{(k)}\right)e^{(k)}$$

6. Repeat Step 5 until it reaches the hidden layer on the immediate right of the input layer.

7. Adjust the weights of the neural network using this learning rule:

$$\Delta w_{ij} = \alpha \delta_i x_j$$
$$w_{ij} \leftarrow w_{ij} + \Delta w_{ij}$$

8. Repeat Steps 4-7 for all training data points.

9. Repeat Steps 4-8 until the neural network has been trained properly.

Although it appears complicated because of its many steps, this process is basically the same as that of the back-propagation of Chapter 3. The detailed explanations are omitted.

Multiclass Classification

This section introduces how to utilize the neural network to deal with the classification of three or more classes. Consider a classification of the given inputs of coordinates (x, y) into one of three classes (see Figure 4-5).

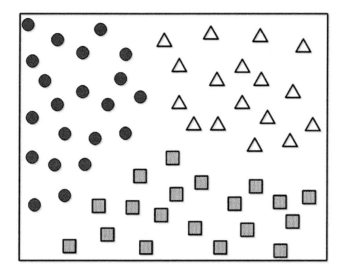

Figure 4-5. *Data with three classes*

We need to construct the neural network first. We will use two nodes for the input layer as the input consists of two parameters. For simplicity, the hidden layers are not considered at this time. We need to determine the number of the output nodes as well. It is widely known that matching the number of output nodes to the number of classes is the most promising method. In this example, we use three output nodes, as the problem requires three classes. Figure 4-6 illustrates the configured neural network.

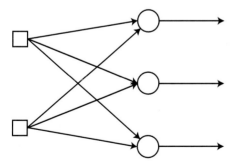

Figure 4-6. *Configured neural network for the three classes*

Once the neural network has been trained with the given data, we obtain the multiclass classifier that we want. The training data is given in Figure 4-7. For each data point, the first two numbers are the x and y coordinates respectively, and the third value is the corresponding class. The data includes the input and correct output as it is used for supervised learning.

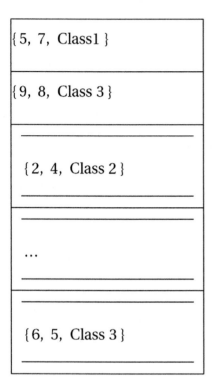

Figure 4-7. Training data with multiclass classifier

In order to calculate the error, we switch the class names into numeric codes, as we did in the previous section. Considering that we have three output nodes from the neural network, we create the classes as the following vectors:

Class 1 → [1 0 0]

Class 2 → [0 1 0]

Class 3 → [0 0 1]

This transformation implies that each output node is mapped to an element of the class vector, which only yields 1 for the corresponding node. For example, if the data belongs to Class 2, the output only yields 1 for the second node and 0 for the others (see Figure 4-8).

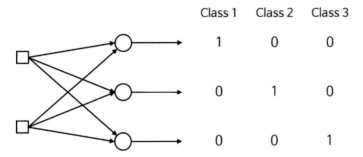

Class 1   Class 2   Class 3

1   0   0

0   1   0

0   0   1

Figure 4-8. Each output node is now mapped to an element of the class vector

This expression technique is called *one-hot encoding* or *1-of-N encoding*. The reason that we match the number of output nodes to the number of classes is to apply this encoding technique. Now, the training data is displayed in the format shown in Figure 4-9.

{ 5, 7, **1, 0, 0** }

{ 9, 8, **0, 0, 1** }

{ 2, 4, **0, 1, 0** }

...

{ 6, 5, **0, 0, 1** }

Figure 4-9. Training data is displayed in a new format

Next, the activation function of the output node should be defined. Since the correct outputs of the transformed training data range from zero to one, can we just use the sigmoid function as we did for the binary classification? In general, multiclass classifiers employ the softmax function as the activation function of the output node.

The activation functions that we have discussed so far, including the sigmoid function, account only for the weighted sum of inputs. They do not consider the output from the other output nodes. However, the softmax function accounts not only for the weighted sum of the inputs, but also for the inputs to the other output nodes. For example, when the weighted sum of the inputs for the three output nodes are 2, 1, and 0.1, respectively, the softmax function calculates the outputs shown in Figure 4-10. All of the weighted sums of the inputs are required in the denominator.

$$
v = \begin{bmatrix} 2 \\ 1 \\ 0.1 \end{bmatrix} \quad \Rightarrow \quad \varphi(v) = \begin{bmatrix} \dfrac{e^2}{e^2 + e^1 + e^{0.1}} \\ \dfrac{e^1}{e^2 + e^1 + e^{0.1}} \\ \dfrac{e^{0.1}}{e^2 + e^1 + e^{0.1}} \end{bmatrix} = \begin{bmatrix} 0.6590 \\ 0.2424 \\ 0.0986 \end{bmatrix}
$$

Figure 4-10. *Softmax function calculations*

Why do we insist on using the softmax function? Consider the sigmoid function in place of the softmax function. Assume that the neural network produced the output shown in Figure 4-11 when given the input data. As the sigmoid function concerns only its own output, the output here will be generated.

Figure 4-11. *Output when using a sigmoid function*

The first output node appears to be in Class 1 by 100 percent probability. Does the data belong to Class 1, then? Not so fast. The other output nodes also indicate 100 percent probability of being in Class 2 and Class 3. Therefore, adequate interpretation of the output from the multiclass classification neural network requires consideration of the relative magnitudes of all node outputs. In this example, the actual probability of being each class is $\frac{1}{3}$. The softmax function provides the correct values.

The softmax function maintains the sum of the output values to be one and also limits the individual outputs to be within the values of 0-1. As it accounts for the relative magnitudes of all the outputs, the softmax function is a suitable choice for the multiclass classification neural networks. The output from the i-th output node of the softmax function is calculated as follows:

$$y_i = \varphi(v_i) = \frac{e^{v_i}}{e^{v_1} + e^{v_2} + e^{v_3} + \cdots + e^{v_M}}$$

$$= \frac{e^{v_i}}{\sum_{k=1}^{M} e^{v_k}}$$

where, v_i is the weighted sum of the i-th output node, and M is the number of output nodes. Following this definition, the softmax function satisfies the following condition:

$$\varphi(v_1) + \varphi(v_2) + \varphi(v_3) + \cdots + \varphi(v_M) = 1$$

Finally, the learning rule should be determined. The multiclass classification neural network usually employs the cross entropy-driven learning rules just like the binary classification network does. This is due to the high learning performance and simplicity that the cross entropy function provides.

Long story short, the learning rule of the multiclass classification neural network is identical to that of the binary classification neural network of the previous section. Although these two neural networks employ different activation functions—the sigmoid for the binary and the softmax for the multiclass—the derivation of the learning rule leads to the same result. Well, it is better for us to have less to remember.

The training process of the multiclass classification neural network is summarized in these steps.

1. Construct the output nodes to have the same value as the number of classes. The softmax function is used as the activation function.

2. Switch the names of the classes into numeric vectors via the one-hot encoding method.

   Class 1 → [1 0 0]

   Class 2 → [0 1 0]

   Class 3 → [0 0 1]

3. Initialize the weights of the neural network with adequate values.

4. Enter the input from the training data { input, correct output } into the neural network and obtain the output. Calculate the error between the output and correct output and determine the delta, δ, of the output nodes.

$$e = d - y$$
$$\delta = e$$

5. Propagate the output delta backwards and calculate the delta of the subsequent hidden nodes.

$$e^{(k)} = W^T \delta$$
$$\delta^{(k)} = \varphi'\left(v^{(k)}\right)e^{(k)}$$

6. Repeat Step 5 until it reaches the hidden layer on the immediate right of the input layer.

7. Adjust the weights of the neural network using this learning rule:

$$\Delta w_{ij} = \alpha \delta_i x_j$$
$$w_{ij} \leftarrow w_{ij} + \Delta w_{ij}$$

8. Repeat Steps 4-7 for all the training data points.

9. Repeat Steps 4-8 until the neural network has been trained properly.

Of course, the multiclass classification neural network is applicable for binary classification. All we have to do is construct a neural network with two output nodes and use the softmax function as the activation function.

Example: Multiclass Classification

In this section, we implement a multiclass classifier network that recognizes digits from the input images. The binary classification has been implemented in Chapter 3, where the input coordinates were divided into two groups. As it classified the data into either 0 or 1, it was binary classification.

Consider an image recognition of digits. This is a multiclass classification, as it classifies the image into specified digits. The input images are five-by-five pixel squares, which display five numbers from 1 to 5, as shown in Figure 4-12.

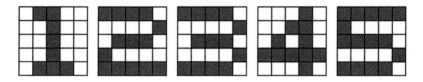

Figure 4-12. *Five-by-five pixel squares that display five numbers from 1 to 5*

The neural network model contains a single hidden layer, as shown in Figure 4-13. As each image is set on a matrix, we set 25 input nodes. In addition, as we have five digits to classify, the network contains five output nodes. The softmax function is used as the activation function of the output node. The hidden layer has 50 nodes and the sigmoid function is used as the activation function.

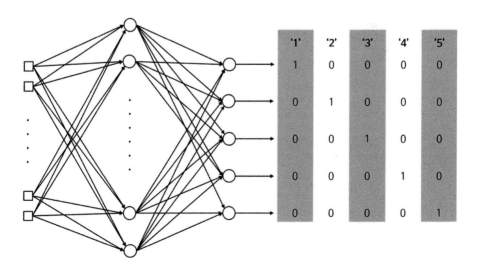

Figure 4-13. *The neural network model for this new dataset*

93

The function MultiClass implements the learning rule of multiclass classification using the SGD method. It takes the input arguments of the weights and training data and returns the trained weights.

```
[W1, W2] = MultiClass(W1, W2, X, D)
```

where W1 and W2 are the weight matrices of the input-hidden and hidden-output layers, respectively. X and D are the input and correct output of the training data, respectively. The following listing shows the MultiClass.m file, which implements the function MultiClass.

```
function [W1, W2] = MultiClass(W1, W2, X, D)
  alpha = 0.9;

  N = 5;
  for k = 1:N
    x = reshape(X(:, :, k), 25, 1);
    d = D(k, :)';

    v1 = W1*x;
    y1 = Sigmoid(v1);
    v  = W2*y1;
    y  = Softmax(v);

    e     = d - y;
    delta = e;

    e1     = W2'*delta;
    delta1 = y1.*(1-y1).*e1;

    dW1 = alpha*delta1*x';
    W1 = W1 + dW1;

    dW2 = alpha*delta*y1';
    W2 = W2 + dW2;
  end
end
```

This code follows the same procedure as that of the example code in the "Cross Entropy Function" section in Chapter 3, which applies the delta rule to the training data, calculates the weight updates, dW1 and dW2, and adjusts the neural network's weights. However, this code slightly differs in that it uses the function softmax for the calculation of the output and calls the function reshape to import the inputs from the training data.

```
x = reshape(X(:, :, k), 25, 1);
```

The input argument X contains the stacked two-dimensional image data. This means that X is a $5\times5\times5$ three-dimensional matrix. Therefore, the first argument of the function reshape, X(:, :, k) indicates the 5×5 matrix that contains the **k**-th image data. As this neural network is compatible with only the vector format inputs, the two-dimensional matrix should be transformed into a 25×1 vector. The function reshape performs this transformation.

Using the cross entropy-driven learning rule, the delta of the output node is calculated as follows:

```
e     = d - y;
delta = e;
```

Similar to the example from Chapter 3, no other calculation is required. This is because, in the cross entropy-driven learning rule that uses the softmax activation function, the delta and error are identical. Of course, the previous back-propagation algorithm applies to the hidden layer.

```
e1     = W2'*delta;
delta1 = y1.*(1-y1).*e1;
```

The function Softmax, which the function MultiClass calls in, is implemented in the Softmax.m file shown in the following listing. This file implements the definition of the softmax function literally. It is simple enough and therefore further explanations have been omitted.

```
function y = Softmax(x)
  ex = exp(x);
  y  = ex / sum(ex);
end
```

The following listing shows the TestMultiClass.m file, which tests the function MultiClass. This program calls MultiClass and trains the neural network 10,000 times. Once the training process has been finished, the program enters the training data into the neural network and displays the output. We can verify the training results via the comparison of the output with the correct output.

```
clear all

rng(3);

X  = zeros(5, 5, 5);

X(:, :, 1) = [ 0 1 1 0 0;
               0 0 1 0 0;
               0 0 1 0 0;
               0 0 1 0 0;
               0 1 1 1 0
             ];

X(:, :, 2) = [ 1 1 1 1 0;
               0 0 0 0 1;
               0 1 1 1 0;
               1 0 0 0 0;
               1 1 1 1 1
             ];

X(:, :, 3) = [ 1 1 1 1 0;
               0 0 0 0 1;
               0 1 1 1 0;
               0 0 0 0 1;
               1 1 1 1 0
             ];

X(:, :, 4) = [ 0 0 0 1 0;
               0 0 1 1 0;
               0 1 0 1 0;
               1 1 1 1 1;
               0 0 0 1 0
             ];

X(:, :, 5) = [ 1 1 1 1 1;
               1 0 0 0 0;
               1 1 1 1 0;
               0 0 0 0 1;
               1 1 1 1 0
             ];
```

```
D = [ 1 0 0 0 0;
      0 1 0 0 0;
      0 0 1 0 0;
      0 0 0 1 0;
      0 0 0 0 1
    ];

W1 = 2*rand(50, 25) - 1;
W2 = 2*rand( 5, 50) - 1;

for epoch = 1:10000            % train
  [W1 W2] = MultiClass(W1, W2, X, D);
end

N = 5;                          % inference
for k = 1:N
  x  = reshape(X(:, :, k), 25, 1);
  v1 = W1*x;
  y1 = Sigmoid(v1);
  v  = W2*y1;
  y  = Softmax(v)
end
```

The input data X of the code is a two-dimensional matrix, which encodes the white pixel into a zero and the black pixel into a unity. For example, the image of the number 1 is encoded in the matrix shown in Figure 4-14.

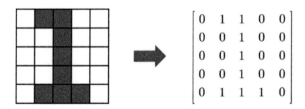

Figure 4-14. *The image of the number 1 is encoded in the matrix*

In contrast, the variable D contains the correct output. For example, the correct output to the first input data, i.e. the image of 1, is located on the first row of the variable D, which is constructed using the one-hot encoding method for each of the five output nodes. Execute the TestMultiClass.m file, and you will see that the neural network has been properly trained in terms of the difference between the output and **D**.

So far, we have verified the neural network for only the training data. However, the practical data does not necessarily reflect the training data. This fact, as we previously discussed, is the fundamental problem of Machine Learning and needs to solve. Let's check our neural network with a simple experiment. Consider the slightly contaminated images shown in Figure 4-15 and watch how the neural network responds to them.

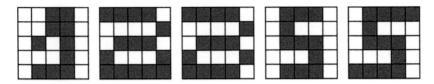

Figure 4-15. *Let's see how the neural network responds to these contaminated images*

The following listing shows the RealMultiClass.m file, which classifies the images shown in Figure 4-15. This program starts with the execution of the TestMultiClass command and trains the neural network. This process yields the weight matrices W1 and W2.

```
clear all

TestMultiClass;                    % W1, W2

X  = zeros(5, 5, 5);

X(:, :, 1) = [ 0 0 1 1 0;
               0 0 1 1 0;
               0 1 0 1 0;
               0 0 0 1 0;
               0 1 1 1 0
             ];

X(:, :, 2) = [ 1 1 1 1 0;
               0 0 0 0 1;
               0 1 1 1 0;
               1 0 0 0 1;
               1 1 1 1 1
             ];
```

```
X(:, :, 3) = [ 1 1 1 1 0;
               0 0 0 0 1;
               0 1 1 1 0;
               1 0 0 0 1;
               1 1 1 1 0
             ];

X(:, :, 4) = [ 0 1 1 1 0;
               0 1 0 0 0;
               0 1 1 1 0;
               0 0 0 1 0;
               0 1 1 1 0
             ];

X(:, :, 5) = [ 0 1 1 1 1;
               0 1 0 0 0;
               0 1 1 1 0;
               0 0 0 1 0;
               1 1 1 1 0
             ];

N = 5;                          % inference
for k = 1:N
  x  = reshape(X(:, :, k), 25, 1);
  v1 = W1*x;
  y1 = Sigmoid(v1);
  v  = W2*y1;
  y  = Softmax(v)
end
```

This code is identical to that of the TestMultiClass.m file, except that it has a different input X and does not include the training process. Execution of this program produces the output of the five contaminated images. Let's take a look one by one.

For the first image, the neural network decided it was a 4 by 96.66% probability. Compare the left and right images in Figure 4-16, which are the input and the digit that the neural network selected, respectively. The input image indeed contains important features of the number 4. Although it appears to be a 1 as well, it is closer to a 4. The classification seems reasonable.

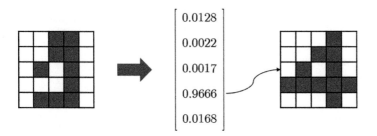

Figure 4-16. *Left and right images are the input and digit that the neural network selected, respectively*

Next, the second image is classified as a 2 by 99.36% probability. This appears to be reasonable when we compare the input image and the training data 2. They only have a one-pixel difference. See Figure 4-17.

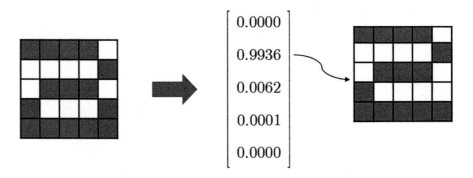

Figure 4-17. *The second image is classified as a 2*

The third image is classified as a 3 by 97.62% probability. This also seems reasonable when we compare the images. See Figure 4-18.

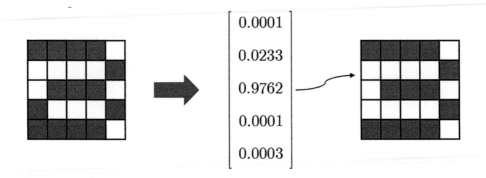

Figure 4-18. *The third image is classified as a 3*

However, when we compare the second and third input images, the difference is only one pixel. This tiny difference results in two totally different classifications. You may not have paid attention, but the training data of these two images has only a two-pixel difference. Isn't it amazing that the neural network catches this small difference and applies it to actual practice?

Let's look at the fourth image. It is classified as a 5 by 47.12% probability. At the same time, it could be a 3 by a pretty high probability of 32.08%. Let's see what happened. The input image appears to be a squeezed 5. Furthermore, the neural network finds some horizontal lines that resemble features of a 3, therefore giving that a high probability. In this case, the neural network should be trained to have more variety in the training data in order to improve its performance.

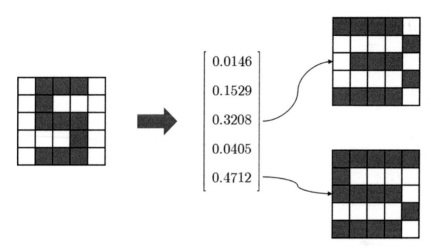

Figure 4-19. *The neural network may have to be trained to have more variety in the training data in order to improve its performance*

Finally, the fifth image is classified as a 5 by 98.18% probability. It is no wonder when we see the input image. However, this image is almost identical to the fourth image. It merely has two additional pixels on the top and bottom of the image. Just extending the horizontal lines results in a dramatic increase in the probability of being a 5. The horizontal feature of a 5 is not as significant in the fourth image. By enforcing this feature, the fifth image is correctly classified as a 5, as shown in Figure 4-20.

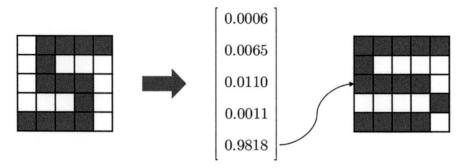

Figure 4-20. *The fifth image is correctly classified as a 5*

Summary

This chapter covered the following concepts:

- For the neural network classifier, the selection of the number of output nodes and activation function usually depends on whether it is for a binary classification (two classes) or for a multiclass classification (three or more classes).

- For binary classification, the neural network is constructed with a single output node and sigmoid activation function. The correct output of the training data is converted to the maximum and minimum values of the activation function. The cost function of the learning rule employs the cross entropy function.

- For a multiclass classification, the neural network includes as many output nodes as the number of classes. The softmax function is employed for the activation function of the output node. The correct output of the training data is converted into a vector using the one-hot encoding method. The cost function of the learning rule employs the cross entropy function.

CHAPTER 5

■ ■ ■

Deep Learning

It's time for Deep Learning. You don't need to be nervous though. As Deep Learning is still an extension of the neural network, most of what you previously read is applicable. Therefore, you don't have many additional concepts to learn.

Briefly, Deep Learning is a Machine Learning technique that employs the deep neural network. As you know, the deep neural network is the multi-layer neural network that contains two or more hidden layers. Although this may be disappointingly simple, this is the true essence of Deep Learning. Figure 5-1 illustrates the concept of Deep Learning and its relationship to Machine Learning.

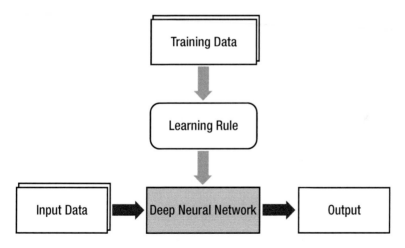

Figure 5-1. *The concept of Deep Learning and its relationship to Machine Learning*

The deep neural network lies in the place of the final product of Machine Learning, and the learning rule becomes the algorithm that generates the model (the deep neural network) from the training data.

© Phil Kim 2017
P. Kim, *MATLAB Deep Learning*, DOI 10.1007/978-1-4842-2845-6_5

Now, knowing that Deep Learning is just the use of a deeper (more hidden layers) neural network, you may ask, "What makes Deep Learning so attractive? Has anyone ever thought of making the neural network's layers even deeper?" In order to answer these questions, we need to look into the history of the neural network.

It did not take very long for the single-layer neural network, the first generation of the neural network, to reveal its fundamental limitations when solving the practical problems that Machine Learning faced.[1] The researchers already knew that the multi-layer neural network would be the next breakthrough. However, it took approximately 30 years until another layer was added to the single-layer neural network. It may not be easy to understand why it took so long for just one additional layer. It was because the proper learning rule for the multi-layer neural network was not found. Since the training is the only way for the neural network to store the information, the untrainable neural network is useless.

The problem of training of the multi-layer neural network was finally solved in 1986 when the back-propagation algorithm was introduced. The neural network was on stage again. However, it was soon met with another problem. Its performance on practical problems did not meet expectations. Of course, there were various attempts to overcome the limitations, including the addition of hidden layers and addition of nodes in the hidden layer. However, none of them worked. Many of them yielded even poorer performances. As the neural network has a very simple architecture and concept, there was nothing much to do that could improve it. Finally, the neural network was sentenced to having no possibility of improvement and it was forgotten.

It remained forgotten for about 20 years until the mid-2000s when Deep Learning was introduced, opening a new door. It took a while for the deep hidden layer to yield sufficient performance because of the difficulties in training the deep neural network. Anyway, the current technologies in Deep Learning yield dazzling levels of performance, which outsmarts the other Machine Learning techniques as well as other neural networks, and prevail in the studies of Artificial Intelligence.

In summary, the reason the multi-layer neural network took 30 years to solve the problems of the single-layer neural network was the lack of the learning rule, which was eventually solved by the back-propagation algorithm. In contrast, the reason another 20 years passed until the introduction of deep neural network-based Deep Learning was the poor performance. The back-propagation training with the additional hidden layers often resulted in poorer performance. Deep Learning provided a solution to this problem.

---

[1]As addressed in Chapter 2, the single-layer neural network can solve only linearly separable problems.

Improvement of the Deep Neural Network

Despite its outstanding achievements, Deep Learning actually does not have any critical technologies to present. The innovation of Deep Learning is a result of many small technical improvements. This section briefly introduces why the deep neural network yielded poor performance and how Deep Learning overcame this problem.

The reason that the neural network with deeper layers yielded poorer performance was that the network was not properly trained. The back-propagation algorithm experiences the following three primary difficulties in the training process of the deep neural network:

- Vanishing gradient

- Overfitting

- Computational load

Vanishing Gradient

The gradient in this context can be thought as a similar concept to the delta of the back-propagation algorithm. The *vanishing gradient* in the training process with the back-propagation algorithm occurs when the output error is more likely to fail to reach the farther nodes. The back-propagation algorithm trains the neural network as it propagates the output error backward to the hidden layers. However, as the error hardly reaches the first hidden layer, the weight cannot be adjusted. Therefore, the hidden layers that are close to the input layer are not properly trained. There is no point of adding hidden layers if they cannot be trained (see Figure 5-2).

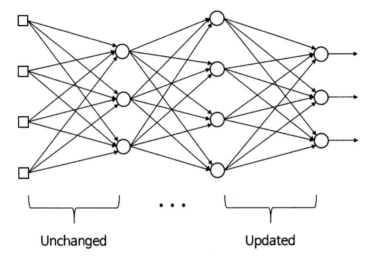

Figure 5-2. *The vanishing gradient*

The representative solution to the vanishing gradient is the use of the **Re**ctified **L**inear **U**nit (ReLU) function as the activation function. It is known to better transmit the error than the sigmoid function. The ReLU function is defined as follows:

$$\varphi(x) = \begin{cases} x, & x > 0 \\ 0, & x \le 0 \end{cases}$$
$$= \max(0, \ x)$$

Figure 5-3 depicts the ReLU function. It produces zero for negative inputs and conveys the input for positive inputs.[2] Its implementation is extremely easy as well.

---

[2]It earned its name as its behavior is similar to that of the rectifier, an electrical element that converts the alternating current into direct current as it cuts out negative voltage.

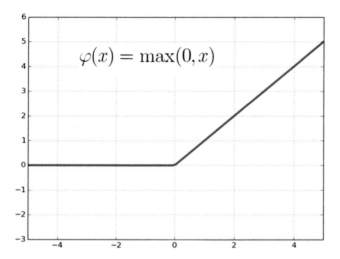

Figure 5-3. *The ReLU function*

The sigmoid function limits the node's outputs to the unity regardless of the input's magnitude. In contrast, the ReLU function does not exert such limits. Isn't it interesting that such a simple change resulted in a drastic improvement of the learning performance of the deep neural network?

Another element that we need for the back-propagation algorithm is the derivative of the ReLU function. By the definition of the ReLU function, its derivative is given as:

$$\varphi'(x) = \begin{cases} 1, & x > 0 \\ 0, & x \le 0 \end{cases}$$

In addition, the cross entropy-driven learning rules may improve the performance, as addressed in Chapter 3. Furthermore, the advanced gradient descent[3], which is a numerical method that better achieves the optimum value, is also beneficial for the training of the deep neural network.

Overfitting

The reason that the deep neural network is especially vulnerable to overfitting is that the model becomes more complicated as it includes more hidden layers, and hence more weight. As addressed in Chapter 1, a complicated model is more vulnerable to overfitting. Here is the dilemma—deepening the layers for

---

[3]sebastianruder.com/optimizing-gradient-descent/

higher performance drives the neural network to face the challenge of Machine Learning.

The most representative solution is the dropout, which trains only some of the randomly selected nodes rather than the entire network. It is very effective, while its implementation is not very complex. Figure 5-4 explains the concept of the dropout. Some nodes are randomly selected at a certain percentage and their outputs are set to be zero to deactivate the nodes.

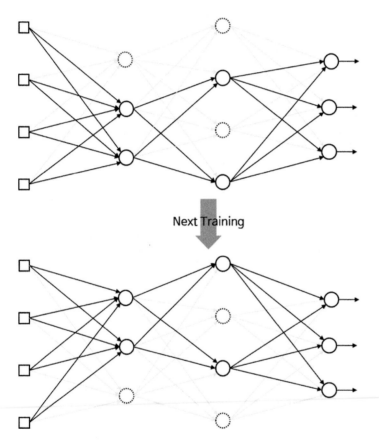

Figure 5-4. *Dropout is where some nodes are randomly selected and their outputs are set to zero to deactivate the nodes*

The dropout effectively prevents overfitting as it continuously alters the nodes and weights in the training process. The adequate percentages of the dropout are approximately 50% and 25% for hidden and input layers, respectively.

Another prevailing method used to prevent overfitting is adding regularization terms, which provide the magnitude of the weights, to the cost function. This method works as it simplifies the neural network' architecture as much as possible, and hence reduces the possible onset of overfitting. Chapter 3 explains this aspect. Furthermore, the use of massive training data is also very helpful as the potential bias due to particular data is reduced.

Computational Load

The last challenge is the time required to complete the training. The number of weights increases geometrically with the number of hidden layers, thus requiring more training data. This ultimately requires more calculations to be made. The more computations the neural network performs, the longer the training takes. This problem is a serious concern in the practical development of the neural network. If a deep neural network requires a month to train, it can only be modified 20 times a year. A useful research study is hardly possible in this situation. This trouble has been relieved to a considerable extent by the introduction of high-performance hardware, such as GPU, and algorithms, such as batch normalization.

The minor improvements that this section introduced are the drivers that has made Deep Learning the hero of Machine Learning. The three primary research areas of Machine Learning are usually said to be the image recognition, speech recognition, and natural language processing. Each of these areas had been separately studied with specifically suitable techniques. However, Deep Learning currently outperforms all the techniques of all three areas.

Example: ReLU and Dropout

This section implements the ReLU activation function and dropout, the representative techniques of Deep Learning. It reuses the example of the digit classification from Chapter 4. The training data is the same five-by-five square images.

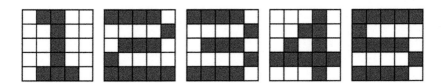

Figure 5-5. *Training data in five-by-five square images*

Consider the deep neural network with the three hidden layers, as shown in Figure 5-6. Each hidden layer contains 20 nodes. The network has 25 input nodes for the matrix input and five output nodes for the five classes. The output nodes employ the softmax activation function.

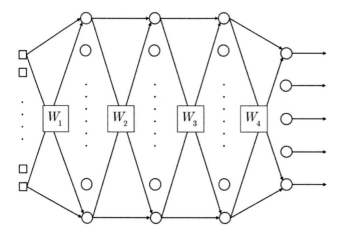

Figure 5-6. *The deep neural network with three hidden layers*

ReLU Function

This section introduces the ReLU function via the example. The function DeepReLU trains the given deep neural network using the back-propagation algorithm. It takes the weights of the network and training data and returns the trained weights.

```
[W1, W2, W3, W4] = DeepReLU(W1, W2, W3, W4, X, D)
```

where W1, W2, W3, and W4 are weight matrices of input-hidden1, hidden1-hidden2, hidden2-hidden3, and hidden3-output layers, respectively. X and D are input and correct output matrices of the training data. The following listing shows the DeepReLU.m file, which implements the DeepReLU function.

```
function [W1, W2, W3, W4] = DeepReLU(W1, W2, W3, W4, X, D)
  alpha = 0.01;

  N = 5;
  for k = 1:N
    x  = reshape(X(:, :, k), 25, 1);
    v1 = W1*x;
    y1 = ReLU(v1);
```

```
      v2 = W2*y1;
      y2 = ReLU(v2);

      v3 = W3*y2;
      y3 = ReLU(v3);

      v  = W4*y3;
      y  = Softmax(v);

      d      = D(k, :)';

      e      = d - y;
      delta = e;

      e3     = W4'*delta;
      delta3 = (v3 > 0).*e3;

      e2     = W3'*delta3;
      delta2 = (v2 > 0).*e2;

      e1     = W2'*delta2;
      delta1 = (v1 > 0).*e1;

      dW4 = alpha*delta*y3';
      W4  = W4 + dW4;

      dW3 = alpha*delta3*y2';
      W3  = W3 + dW3;

      dW2 = alpha*delta2*y1';
      W2  = W2 + dW2;

      dW1 = alpha*delta1*x';
      W1  = W1 + dW1;
  end
end
```

This code imports the training data, calculates the weight updates (dW1, dW2, dW3, and dW4) using the delta rule, and adjusts the weight of the neural network. So far, the process is identical to the previous training codes. It only differs in that the hidden nodes employ the function ReLU, in place of sigmoid. Of course, the use of a different activation function yields a change in its derivative as well.

Now, let's look into the function ReLU that the function DeepReLU calls. The listing of the function ReLU shown here is implemented in the ReLU.m file. As this is just a definition, further discussion is omitted.

```
function y = ReLU(x)
  y = max(0, x);
end
```

Consider the back-propagation algorithm portion, which adjusts the weights using the back-propagation algorithm. The following listing shows the extract of the delta calculation from the DeepReLU.m file. This process starts from the delta of the output node, calculates the error of the hidden node, and uses it for the next error. It repeats the same steps through delta3, delta2, and delta1.

```
...
e     = d - y;
delta = e;

e3     = W4'*delta;
delta3 = (v3 > 0).*e3;

e2     = W3'*delta3;
delta2 = (v2 > 0).*e2;

e1     = W2'*delta2;
delta1 = (v1 > 0).*e1;
...
```

Something noticeable from the code is the derivative of the function ReLU. For example, in the calculation of the delta of the third hidden layer, delta3, the derivative of the ReLU function is coded as follows:

```
(v3 > 0)
```

Let's see how this line becomes the derivative of the ReLU function. MATLAB returns a unity and zero if the expressions in the brackets are true and false, respectively. Therefore, this line becomes 1 if v3 > 0 and 0 otherwise. The same result is produced as the definition of the derivative of the ReLU function shown here:

$$\varphi'(x) = \begin{cases} 1, & x > 0 \\ 0, & x \leq 0 \end{cases}$$

The following listing shows the TestDeepReLU.m file, which tests the DeepReLU function. This program calls the DeepReLU function and trains the network 10,000 times. It enters the training data into the trained network and displays the output. We verify the adequacy of the training by comparing the output and correct output.

```
clear all

X  = zeros(5, 5, 5);

X(:, :, 1) = [ 0 1 1 0 0;
               0 0 1 0 0;
               0 0 1 0 0;
               0 0 1 0 0;
               0 1 1 1 0
            ];

X(:, :, 2) = [ 1 1 1 1 0;
               0 0 0 0 1;
               0 1 1 1 0;
               1 0 0 0 0;
               1 1 1 1 1
            ];

X(:, :, 3) = [ 1 1 1 1 0;
               0 0 0 0 1;
               0 1 1 1 0;
               0 0 0 0 1;
               1 1 1 1 0
            ];

X(:, :, 4) = [ 0 0 0 1 0;
               0 0 1 1 0;
               0 1 0 1 0;
               1 1 1 1 1;
               0 0 0 1 0
            ];

X(:, :, 5) = [ 1 1 1 1 1;
               1 0 0 0 0;
               1 1 1 1 0;
               0 0 0 0 1;
               1 1 1 1 0
            ];

D = [ 1 0 0 0 0;
      0 1 0 0 0;
      0 0 1 0 0;
      0 0 0 1 0;
      0 0 0 0 1
    ];
```

```
W1 = 2*rand(20, 25) - 1;
W2 = 2*rand(20, 20) - 1;
W3 = 2*rand(20, 20) - 1;
W4 = 2*rand( 5, 20) - 1;

for epoch = 1:10000              % train
  [W1, W2, W3, W4] = DeepReLU(W1, W2, W3, W4, X, D);
end

N = 5;                           % inference
for k = 1:N
  x  = reshape(X(:, :, k), 25, 1);
  v1 = W1*x;
  y1 = ReLU(v1);

  v2 = W2*y1;
  y2 = ReLU(v2);

  v3 = W3*y2;
  y3 = ReLU(v3);

  v  = W4*y3;
  y  = Softmax(v)
end
```

As this code is also almost identical to the previous test programs, a detailed explanation is omitted. This code occasionally fails to train properly and yields wrong outputs, which has never happened with the sigmoid activation function. The sensitivity of the ReLU function to the initial weight values seems to cause this anomaly.

Dropout

This section presents the code that implements the dropout. We use the sigmoid activation function for the hidden nodes. This code is mainly used to see how the dropout is coded, as the training data may be too simple for us to perceive the actual improvement of overfitting.

The function DeepDropout trains the example neural network using the back-propagation algorithm. It takes the neural network's weights and training data and returns the trained weights.

```
[W1, W2, W3, W4] = DeepDropout(W1, W2, W3, W4, X, D)
```

where the notation of the variables is the same as that of the function DeepReLU of the previous section. The following listing shows the DeepDropout.m file, which implements the DeepDropout function.

```
function [W1, W2, W3, W4] = DeepDropout(W1, W2, W3, W4, X, D)
  alpha = 0.01;

  N = 5;
  for k = 1:N
    x  = reshape(X(:, :, k), 25, 1);
    v1 = W1*x;
    y1 = Sigmoid(v1);
    y1 = y1 .* Dropout(y1, 0.2);

    v2 = W2*y1;
    y2 = Sigmoid(v2);
    y2 = y2 .* Dropout(y2, 0.2);

    v3 = W3*y2;
    y3 = Sigmoid(v3);
    y3 = y3 .* Dropout(y3, 0.2);

    v  = W4*y3;
    y  = Softmax(v);

    d      = D(k, :)';

    e      = d - y;
    delta = e;

    e3      = W4'*delta;
    delta3 = y3.*(1-y3).*e3;

    e2      = W3'*delta3;
    delta2 = y2.*(1-y2).*e2;

    e1      = W2'*delta2;
    delta1 = y1.*(1-y1).*e1;

    dW4 = alpha*delta*y3';
    W4  = W4 + dW4;

    dW3 = alpha*delta3*y2';
    W3  = W3 + dW3;
```

```
    dW2 = alpha*delta2*y1';
    W2  = W2 + dW2;

    dW1 = alpha*delta1*x';
    W1  = W1 + dW1;
  end
end
```

This code imports the training data, calculates the weight updates (dW1, dW2, dW3, and dW4) using the delta rule, and adjusts the weight of the neural network. This process is identical to that of the previous training codes. It differs from the previous ones in that once the output is calculated from the Sigmoid activation function of the hidden node, the Dropout function modifies the final output of the node. For example, the output of the first hidden layer is calculated as:

```
y1 = Sigmoid(v1);
y1 = y1 .* Dropout(y1, 0.2);
```

Executing these lines switches the outputs from 20% of the first hidden nodes to 0; it drops out 20% of the first hidden nodes.

Here are the details of the implementation of the function Dropout. It takes the output vector and dropout ratio and returns the new vector that will be multiplied to the output vector.

```
ym = Dropout(y, ratio)
```

where y is the output vector and ratio is the ratio of the dropout of the output vector. The return vector ym of the function Dropout has the same dimensions as y. ym contains zeros for as many elements as the ratio and $1/(1-ratio)$ for the other elements. Consider the following example:

```
y1 = rand(6, 1)
ym = Dropout(y1, 0.5)
y1 = y1 .* ym
```

The function Dropout implements the dropout. Executing this code will display the results shown in Figure 5-7.

$$
y_1 = \begin{bmatrix} 0.5356 \\ 0.9537 \\ 0.5442 \\ 0.0821 \\ 0.3663 \\ 0.8509 \end{bmatrix} \quad ym = \begin{bmatrix} 2 \\ 2 \\ 0 \\ 0 \\ 0 \\ 2 \end{bmatrix} \quad y_1 * ym = \begin{bmatrix} 1.0712 \\ 1.9075 \\ 0 \\ 0 \\ 0 \\ 1.7017 \end{bmatrix}
$$

Figure 5-7. *The dropout function in action*

The vector ym has three elements: half (0.5) of the six elements of the vector y1, which are filled with zeroes, and the others are filled with $1/(1-0.5)$, which equals 2. When this ym is multiplied to the original vector y1, the revised y1 has zeros by the specified ratio. In other words, y1 drops out the specified portion of the elements.

The reason that we multiply the other element by $1/(1-ratio)$ is to compensate for the loss of output due to the dropped elements. In the previous example, once half of the vector y1 has been dropped out, the magnitude of the layer's output significantly diminishes. Therefore, the outputs of the survived nodes are amplified by the proper proportion.

The function Dropout is implemented in the Dropout.m file:

```
function ym = Dropout(y, ratio)
  [m, n] = size(y);
  ym     = zeros(m, n);

  num      = round(m*n*(1-ratio));
  idx      = randperm(m*n, num);
  ym(idx) = 1 / (1-ratio);
end
```

The explanation is long, but the code itself is very simple. The code prepares the zero matrix ym, of which the dimension is the same as that of y. It calculates the number of survivors, num, based on the given dropout ratio, ratio, and randomly selects the survivors from ym. Specifically, it selects the indices of the elements of ym. This is done by the randperm portion of the code. Now that the code has the indices of the non-zero elements, put $1/(1-ratio)$ into those elements. The other elements are already filled with zeros, as the vector ym has been a zero matrix from the beginning.

The following listing shows the TestDeepDropout.m file, which tests the DeepDropout function. This program calls DeepDropout and trains the neural network 20,000 times. It enters the training data into the trained network and

displays the output. We verify the adequacy of the training by comparing the
output and correct output.

```
clear all

X  = zeros(5, 5, 5);

X(:, :, 1) = [ 0 1 1 0 0;
               0 0 1 0 0;
               0 0 1 0 0;
               0 0 1 0 0;
               0 1 1 1 0
             ];

X(:, :, 2) = [ 1 1 1 1 0;
               0 0 0 0 1;
               0 1 1 1 0;
               1 0 0 0 0;
               1 1 1 1 1
             ];

X(:, :, 3) = [ 1 1 1 1 0;
               0 0 0 0 1;
               0 1 1 1 0;
               0 0 0 0 1;
               1 1 1 1 0
             ];

X(:, :, 4) = [ 0 0 0 1 0;
               0 0 1 1 0;
               0 1 0 1 0;
               1 1 1 1 1;
               0 0 0 1 0
             ];

X(:, :, 5) = [ 1 1 1 1 1;
               1 0 0 0 0;
               1 1 1 1 0;
               0 0 0 0 1;
               1 1 1 1 0
             ];
```

```
D = [ 1 0 0 0 0;
      0 1 0 0 0;
      0 0 1 0 0;
      0 0 0 1 0;
      0 0 0 0 1
    ];

W1 = 2*rand(20, 25) - 1;
W2 = 2*rand(20, 20) - 1;
W3 = 2*rand(20, 20) - 1;
W4 = 2*rand( 5, 20) - 1;

for epoch = 1:20000          % train
  [W1, W2, W3, W4] = DeepDropout(W1, W2, W3, W4, X, D);
end

N = 5;                       % inference
for k = 1:N
  x  = reshape(X(:, :, k), 25, 1);
  v1 = W1*x;
  y1 = Sigmoid(v1);

  v2 = W2*y1;
  y2 = Sigmoid(v2);

  v3 = W3*y2;
  y3 = Sigmoid(v3);

  v  = W4*y3;
  y  = Softmax(v)
end
```

This code is almost identical to the other test codes. The only difference is that it calls the DeepDropout function when it calculates the output of the trained network. Further explanation is omitted.

Summary

This chapter covered the following topics:

- Deep Learning can be simply defined as a Machine Learning technique that employs the deep neural network.

- The previous neural networks had a problem where the deeper (more) hidden layers were harder to train and degraded the performance. Deep Learning solved this problem.

- The outstanding achievements of Deep Learning were not made by a critical technique but rather are due to many minor improvements.

- The poor performance of the deep neural network is due to the failure of proper training. There are three major showstoppers: the vanishing gradient, overfitting, and computational load.

- The vanishing gradient problem is greatly improved by employing the ReLU activation function and the cross entropy-driven learning rule. Use of the advanced gradient descent method is also beneficial.

- The deep neural network is more vulnerable to overfitting. Deep Learning solves this problem using the dropout or regularization.

- The significant training time is required due to the heavy calculations. This is relieved to a large extent by the GPU and various algorithms.

CHAPTER 6

■ ■ ■

Convolutional Neural Network

Chapter 5 showed that incomplete training is the cause of the poor performance of the deep neural network and introduced how Deep Learning solved the problem. The importance of the deep neural network lies in the fact that it opened the door to the complicated non-linear model and systematic approach for the hierarchical processing of knowledge.

This chapter introduces the convolutional neural network (ConvNet), which is a deep neural network specialized for image recognition. This technique exemplifies how significant the improvement of the deep layers is for information (images) processing. Actually, ConvNet is an old technique, which was developed in the 1980s and 1990s.[1] However, it has been forgotten for a while, as it was impractical for real-world applications with complicated images. Since 2012 when it was dramatically revived[2], ConvNet has conquered most computer vision fields and is growing at a rapid pace.

Architecture of ConvNet

ConvNet is not just a deep neural network that has many hidden layers. It is a deep network that imitates how the visual cortex of the brain processes and recognizes images. Therefore, even the experts of neural networks often have a hard time understanding this concept on their first encounter. That is how much ConvNet differs in concept and operation from the previous neural networks. This section briefly introduces the fundamental architecture of ConvNet.

---

[1]LeCun, Y., et al., "Handwritten digit recognition with a back-propagation network," In *Proc. Advances in Neural Information Processing Systems*, 396–404 (1990).
[2]Krizhevsky, Alex, "ImageNet Classification with Deep Convolutional Neural Networks," 17 November 2013.

© Phil Kim 2017
P. Kim, *MATLAB Deep Learning*, DOI 10.1007/978-1-4842-2845-6_6

Basically, image recognition is the classification. For example, recognizing whether the image of a picture is a cat or a dog is the same as classifying the image into a cat or dog class. The same thing applies to the letter recognition; recognizing the letter from an image is the same as classifying the image into one of the letter classes. Therefore, the output layer of the ConvNet generally employs the multiclass classification neural network.

However, directly using the original images for image recognition leads to poor results, regardless of the recognition method; the images should be processed to contrast the features. The examples in Chapter 4 used the original images and they worked well because they were simple black-and-white images. Otherwise, the recognition process would have ended up with very poor results. For this reason, various techniques for image feature extraction have been developed.[3]

Before ConvNet, the feature extractor has been designed by experts of specific areas. Therefore, it required a significant amount of cost and time while it yielded an inconsistent level of performance. These feature extractors were independent of Machine Learning. Figure 6-1 illustrates this process.

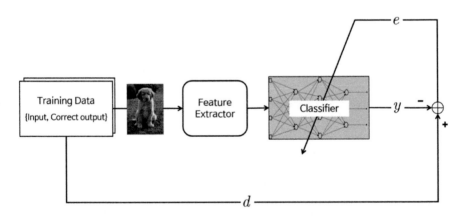

Figure 6-1. *Feature extractors used to be independent of Machine Learning*

ConvNet includes the feature extractor in the training process rather than designing it manually. The feature extractor of ConvNet is composed of special kinds of neural networks, of which the weights are determined via the training process. The fact that ConvNet turned the manual feature extraction design into the automated process is its primary feature and advantage. Figure 6-2 depicts the training concept of ConvNet.

---

[3]The representative methods include SIFT, HoG, Textons, Spin image, RIFT, and GLOH.

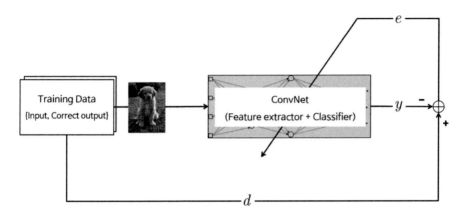

Figure 6-2. *ConvNet's feature extractor is composed of special kinds of neural networks*

ConvNet yields better image recognition when its feature extraction neural network is deeper (contains more layers), at the cost of difficulties in the training process, which had driven ConvNet to be impractical and forgotten for a while.

Let's go a bit deeper. ConvNet consists of a neural network that extracts features of the input image and another neural network that classifies the feature image. Figure 6-3 shows the typical architecture of ConvNet.

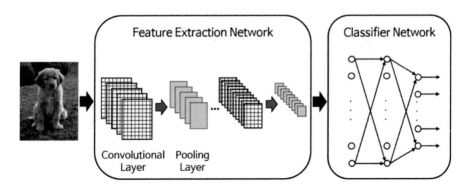

Figure 6-3. *Typical architecture of ConvNet*

The input image enters into the feature extraction network. The extracted feature signals enter the classification neural network. The classification neural network then operates based on the features of the image and generates the output. The classification techniques discussed in Chapter 4 apply here.

The feature extraction neural network consists of piles of the convolutional layer and pooling layer pairs. The convolution layer, as its name implies, converts the image using the convolution operation. It can be thought of as a collection of digital filters. The pooling layer combines the neighboring pixels into a single pixel. Therefore, the pooling layer reduces the dimension of the image. As the primary concern of ConvNet is the image; the operations of the convolution and pooling layers are conceptually in a two-dimensional plane. This is one of the differences between ConvNet and other neural networks.

In summary, ConvNet consists of the serial connection of the feature extraction network and the classification network. Through the training process, the weights of both layers are determined. The feature extraction layer has piled pairs of the convolution and pooling layers. The convolution layer converts the images via the convolution operation, and the pooling layer reduces the dimension of the image. The classification network usually employs the ordinary multiclass classification neural network.

Convolution Layer

This section explains how the convolution layer, which is one side of the feature extraction neural network, works. The pooling layer, the other side of the pair, is introduced in the next section.

The convolution layer generates new images called *feature maps*. The feature map accentuates the unique features of the original image. The convolution layer operates in a very different way compared to the other neural network layers. This layer does not employ connection weights and a weighted sum.[4] Instead, it contains filters[5] that convert images. We will call these filters *convolution filters*. The process of the inputting the image through the convolution filters yields the feature map.

Figure 6-4 shows the process of the convolution layer, where the circled * mark denotes the convolution operation, and the φ mark is the activation function. The square grayscale icons between these operators indicate the convolution filters. The convolution layer generates the same number of feature maps as the convolution filters. Therefore, for instance, if the convolution layer contains four filters, it will generate four feature maps.

---

[4]It is often explained using the local receptive filed and shared weights from the perspective of the ordinary neural network. However, they would not be helpful for beginners. This book does not insist its relationship with the ordinary neural network and explains it as a type of digital filter.
[5]Also called kernels.

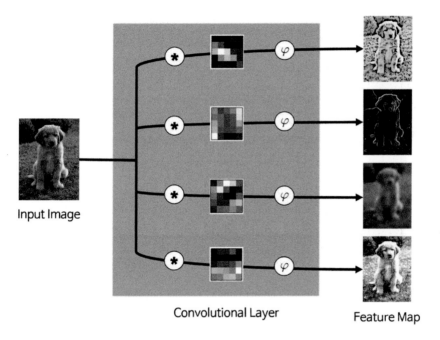

Figure 6-4. *The convolution layer process*

Let's further explore the details of the convolution filter. The filters of the convolution layer are two-dimensional matrices. They usually come in 5×5 or 3×3 matrices, and even 1×1 convolution filters have been used in recent applications. Figure 6-4 shows the values of the 5×5 filters in grayscale pixels. As addressed in the previous section, the values of the filter matrix are determined through the training process. Therefore, these values are continuously trained throughout the training process. This aspect is similar to the updating process of the connection weights of the ordinary neural network.

The *convolution* is a rather difficult operation to explain in text as it lies on the two-dimensional plane. However, its concept and calculation steps are simpler than they appear.[6] A simple example will help you understand how it works. Consider a 4×4 pixel image that is expressed as the matrix shown in Figure 6-5. We will generate a feature map via the convolution filter operation of this image.

---

[6]deeplearning.stanford.edu/wiki/images/6/6c/Convolution_schematic.gif

1	1	1	3
4	6	4	8
30	0	1	5
0	2	2	4

Figure 6-5. *Four-by-four pixel image*

We use the two convolution filters shown here. It should be noted that the filters of the actual ConvNet are determined through the training process and not by manual decision.

$$\begin{bmatrix} 1 & 0 \\ 0 & 1 \end{bmatrix}, \begin{bmatrix} 0 & 1 \\ 1 & 0 \end{bmatrix}$$

Let's start with the first filter. The convolution operation begins at the upper-left corner of the submatrix that is the same size as the convolution filter (see Figure 6-6).

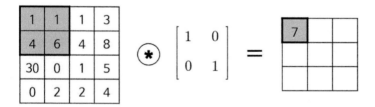

Figure 6-6. *The convolution operation starts at the upper-left corner*

The convolution operation is the sum of the products of the elements that are located on the same positions of the two matrices. The result of 7 in Figure 6-6 is calculated as:

$$(1\times1) + (1\times0) + (4\times0) + (6\times1) \ = \ 7$$

Another convolution operation is conducted for the next submatrix (see Figure 6-7).[7]

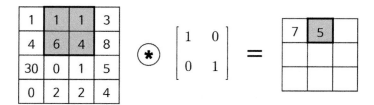

Figure 6-7. *The second convolution operation*

In the same manner, the third convolution operation is conducted, as shown in Figure 6-8.

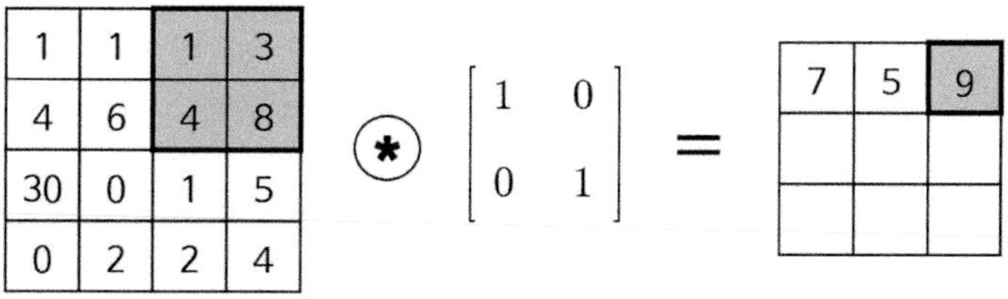

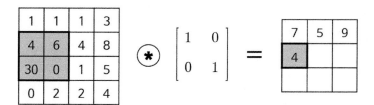

Figure 6-8. *The third convolution operation*

Once the top row is finished, the next row starts over from the left (see Figure 6-9).

Figure 6-9. *The convolution operation starts over from the left*

---

[7]The designer decides how many elements to stride for each operation. It can be greater than one if the filter is larger.

It repeats the same process until the feature map of the given filter is produced, as shown in Figure 6-10.

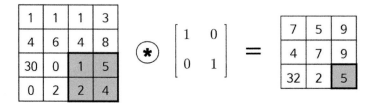

Figure 6-10. *The feature map of the given filter has been completed*

Now, take a closer look at the feature map. The element of (3, 1) of the map shows the greatest value. What happened to this cell? This value is the result of the convolution operation shown in Figure 6-11.

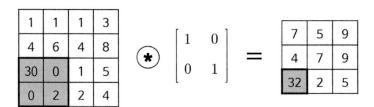

Figure 6-11. *The submatrix of the image matches the convolution filter*

It is noticeable from the figure that the submatrix of the image matches the convolution filter; both are diagonal matrices with significant numbers on the same cells. The convolution operation yields large values when the input matches the filter, as shown in Figure 6-12.

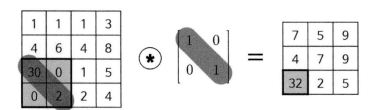

Figure 6-12. *The convolution operation yields large values when the input matches the filter*

In contrast, in the case shown in Figure 6-13, the same significant number of 30 does not affect the convolution result, which is only 4. This is because the image matrix does not match the filter; the significant elements of the image matrix are aligned in the wrong direction.

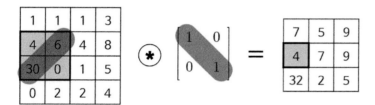

Figure 6-13. *When the image matrix does not match the filter, the significant elements are not aligned*

In the same manner, processing the second convolution filter produces the feature map shown in Figure 6-14.

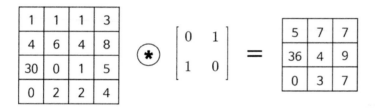

Figure 6-14. *The values depend on whether the image matrix matches the convolution filter*

Similarly to the first convolution operation, the values in the elements of this feature map depend on whether the image matrix matches the convolution filter or not.

In summary, the convolution layer operates the convolution filters on the input image and produces the feature maps. The features that are extracted in the convolution layer determined by the trained convolution filters. Therefore, the features that the convolution layer extracts vary depending on which convolution filter is used.

The feature map that the convolution filter creates is processed through the activation function before the layer yields the output. The activation function of the convolution layer is identical to that of the ordinary neural network. Although the ReLU function is used in most of the recent applications, the sigmoid function and the tanh function are often employed as well.[8]

Just for the reference, the moving average filter, which is widely used in the digital signal processing field, is a special type of convolution filter. If you are familiar with digital filters, relating them to this concept may allow you to better understand the ideas behind the convolution filter.

Pooling Layer

The *pooling layer* reduces the size of the image, as it combines neighboring pixels of a certain area of the image into a single representative value. Pooling is a typical technique that many other image processing schemes have already been employing.

In order to conduct the operations in the pooling layer, we should determine how to select the pooling pixels from the image and how to set the representative value. The neighboring pixels are usually selected from the square matrix, and the number of pixels that are combined differs from problem to problem. The representative value is usually set as the mean or maximum of the selected pixels.

The operation of the pooling layer is surprisingly simple. As it is a two-dimensional operation, and an explanation in text may lead to more confusion, let's go through an example. Consider the 4×4 pixel input image, which is expressed by the matrix shown in Figure 6-15.

1	1	1	3
4	6	4	8
30	0	1	5
0	2	2	4

Figure 6-15. *The four-by-four pixel input image*

We combine the pixels of the input image into a 2×2 matrix without overlapping the elements. Once the input image passes through the pooling layer, it shrinks into a 2×2 pixel image. Figure 6-16 shows the resultant cases of pooling using the mean pooling and max pooling.

---

[8]Sometimes the activation function is omitted depending on the problem.

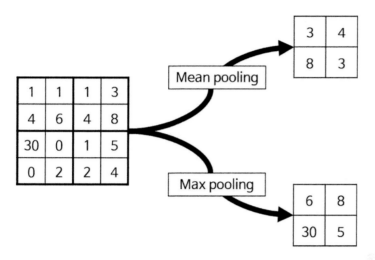

Figure 6-16. *The resultant cases of pooling using two different methods*

Actually, in a mathematical sense, the pooling process is a type of convolution operation. The difference from the convolution layer is that the convolution filter is stationary, and the convolution areas do not overlap. The example provided in the next section will elaborate on this.

The pooling layer compensates for eccentric and tilted objects to some extent. For example, the pooling layer can improve the recognition of a cat, which may be off-center in the input image. In addition, as the pooling process reduces the image size, it is highly beneficial for relieving the computational load and preventing overfitting.

Example: MNIST

We implement a neural network that takes the input image and recognizes the digit that it represents. The training data is the MNIST[9] database, which contains 70,000 images of handwritten numbers. In general, 60,000 images are used for training, and the remaining 10,000 images are used for the validation test. Each digit image is a 28-by-28 pixel black-and-white image, as shown in Figure 6-17.

---

[9]Mixed National Institute of Standards and Technology.

Figure 6-17. *A 28-by-28 pixel black-and-white image from the MNIST database*

Considering the training time, this example employs only 10,000 images with the training data and verification data in an 8:2 ratio. Therefore, we have 8,000 MNIST images for training and 2,000 images for validation of the performance of the neural network. As you may know well by now, the MNIST problem is caused by the multiclass classification of the 28×28 pixel image into one of the ten digit classes of 0-9.

Let's consider a ConvNet that recognizes the MNIST images. As the input is a 28×28 pixel black-and-white image, we allow 784(=28x28) input nodes. The feature extraction network contains a single convolution layer with 20 9×9 convolution filters. The output from the convolution layer passes through the ReLU function, followed by the pooling layer. The pooling layer employs the mean pooling process of two by two submatrices. The classification neural network consists of the single hidden layer and output layer. This hidden layer has 100 nodes that use the ReLU activation function. Since we have 10 classes to classify, the output layer is constructed with 10 nodes. We use the softmax activation function for the output nodes. The following table summarizes the example neural network.

Layer	Remark	Activation Function
Input	28×28 nodes	-
Convolution	20 convolution filters (9×9)	ReLU
Pooling	1 mean pooling (2×2)	-
Hidden	100 nodes	ReLU
Output	10 nodes	Softmax

Figure 6-18 shows the architecture of this neural network. Although it has many layers, only three of them contain the weight matrices that require training; they are W_1, W_5, and W_o in the square blocks. W_5 and W_o contain the connection weights of the classification neural network, while W_1 is the convolution layer's weight, which is used by the convolution filters for image processing.

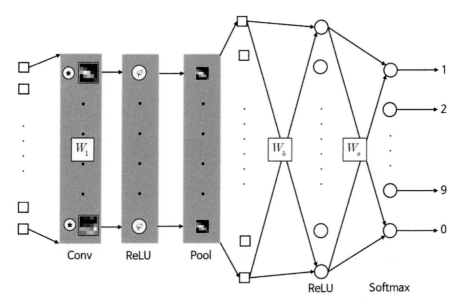

Figure 6-18. *The architecture of this neural network*

The input nodes between the pooling layer and the hidden layer, which are the square nodes left of the W_5 block, are the transformations of the two-dimensional image into a vector. As this layer does not involve any operations, these nodes are denoted as squares.

The function MnistConv, which trains the network using the back-propagation algorithm, takes the neural network's weights and training data and returns the trained weights.

```
[W1, W5, Wo] = MnistConv(W1, W5, Wo, X, D)
```

where W1, W5, and Wo are the convolution filter matrix, pooling-hidden layer weight matrix, and hidden-output layer weight matrix, respectively. X and D are the input and correct output from the training data, respectively. The following listing shows the MnistConv.m file, which implements the MnistConv function.

```
function [W1, W5, Wo] = MnistConv(W1, W5, Wo, X, D)
%
%
```

```
alpha = 0.01;
beta  = 0.95;

momentum1 = zeros(size(W1));
momentum5 = zeros(size(W5));
momentumo = zeros(size(Wo));

N = length(D);

bsize = 100;
blist = 1:bsize:(N-bsize+1);

% One epoch loop
%
for batch = 1:length(blist)
  dW1 = zeros(size(W1));
  dW5 = zeros(size(W5));
  dWo = zeros(size(Wo));

  % Mini-batch loop
  %
  begin = blist(batch);
  for k = begin:begin+bsize-1
    % Forward pass = inference
    %
    x  = X(:, :, k);          % Input,         28x28
    y1 = Conv(x, W1);         % Convolution,   20x20x20
    y2 = ReLU(y1);            %
    y3 = Pool(y2);            % Pool,          10x10x20
    y4 = reshape(y3, [], 1);  %                2000
    v5 = W5*y4;               % ReLU,          360
    y5 = ReLU(v5);            %
    v  = Wo*y5;               % Softmax,       10
    y  = Softmax(v);          %

    % One-hot encoding
    %
    d = zeros(10, 1);
    d(sub2ind(size(d), D(k), 1)) = 1;

    % Backpropagation
    %
    e     = d - y;            % Output layer
    delta = e;
```

```
    e5     = Wo' * delta;               % Hidden(ReLU) layer
    delta5 = (y5 > 0) .* e5;

    e4     = W5' * delta5;              % Pooling layer

    e3     = reshape(e4, size(y3));

    e2 = zeros(size(y2));
    W3 = ones(size(y2)) / (2*2);
    for c = 1:20
      e2(:, :, c) = kron(e3(:, :, c), ones([2 2])) .* W3(:, :, c);
    end

    delta2 = (y2 > 0) .* e2;           % ReLU layer

    delta1_x = zeros(size(W1));        % Convolutional layer
    for c = 1:20
      delta1_x(:, :, c) = conv2(x(:, :), rot90(delta2(:, :, c), 2),
'valid');
    end

    dW1 = dW1 + delta1_x;
    dW5 = dW5 + delta5*y4';
    dWo = dWo + delta *y5';
  end

  % Update weights
  %
  dW1 = dW1 / bsize;
  dW5 = dW5 / bsize;
  dWo = dWo / bsize;

  momentum1 = alpha*dW1 + beta*momentum1;
  W1        = W1 + momentum1;

  momentum5 = alpha*dW5 + beta*momentum5;
  W5        = W5 + momentum5;

  momentumo = alpha*dWo + beta*momentumo;
  Wo        = Wo + momentumo;
end

end
```

This code appears to be rather more complex than the previous examples. Let's take a look at it part by part. The function MnistConv trains the network via the minibatch method, while the previous examples employed the SGD and batch methods. The minibatch portion of the code is extracted and shown in the following listing.

```
bsize = 100;
blist = 1:bsize:(N-bsize+1);

for batch = 1:length(blist)
  ...
  begin = blist(batch);
  for k = begin:begin+bsize-1
    ...
    dW1 = dW1 + delta2_x;
    dW5 = dW5 + delta5*y4';
    dWo = dWo + delta *y5';
  end
  dW1 = dW1 / bsize;
  dW5 = dW5 / bsize;
  dWo = dWo / bsize;
  ...
end
```

The number of batches, bsize, is set to be 100. As we have a total 8,000 training data points, the weights are adjusted 80 (=8,000/100) times for every epoch. The variable blist contains the location of the first training data point to be brought into the minibatch. Starting from this location, the code brings in 100 data points and forms the training data for the minibatch. In this example, the variable blist stores the following values:

```
blist = [ 1, 101, 201, 301, ..., 7801, 7901 ]
```

Once the starting point, begin, of the minibatch is found via blist, the weight update is calculated for every 100th data point. The 100 weight updates are summed and averaged, and the weights are adjusted. Repeating this process 80 times completes one epoch.

Another noticeable aspect of the function MnistConv is that it adjusts the weights using momentum. The variables momentum1, momentum5, and momentumo are used here. The following part of the code implements the momentum update:

```
...
momentum1 = alpha*dW1 + beta*momentum1;
W1        = W1 + momentum1;

momentum5 = alpha*dW5 + beta*momentum5;
W5        = W5 + momentum5;

momentumo = alpha*dWo + beta*momentumo;
Wo        = Wo + momentumo;
...
```

We have now captured the big picture of the code. Now, let's look at the learning rule, the most important part of the code. The process itself is not distinct from the previous ones, as ConvNet also employs back-propagation training. The first thing that must be obtained is the output of the network. The following listing shows the output calculation portion of the function MnistConv. It can be intuitively understood from the architecture of the neural network. The variable y of this code is the final output of the network.

```
...
x  = X(:, :, k);            % Input,                28x28
y1 = Conv(x, W1);           % Convolution,   20x20x20
y2 = ReLU(y1);              %
y3 = Pool(y2);              % Pool,          10x10x20
y4 = reshape(y3, [], 1);    %                         2000
v5 = W5*y4;                 % ReLU,                    360
y5 = ReLU(v5);             %
v  = Wo*y5;                 % Softmax,                  10
y  = Softmax(v);           %
...
```

Now that we have the output, the error can be calculated. As the network has 10 output nodes, the correct output should be in a 10×1 vector in order to calculate the error. However, the MNIST data gives the correct output as the respective digit. For example, if the input image indicates a 4, the correct output will be given as a 4. The following listing converts the numerical correct output into a 10×1 vector. Further explanation is omitted.

```
d = zeros(10, 1);
d(sub2ind(size(d), D(k), 1)) = 1;
```

The last part of the process is the back-propagation of the error. The following listing shows the back-propagation from the output layer to the subsequent layer to the pooling layer. As this example employs cross entropy and softmax functions, the output node delta is the same as that of the network output error. The next hidden layer employs the ReLU activation function. There is nothing particular there. The connecting layer between the hidden and pooling layers is just a rearrangement of the signal.

```
...
e       = d - y;
delta   = e;

e5      = Wo' * delta;
delta5 = e5 .* (y5> 0);

e4      = W5' * delta5;
e3      = reshape(e4, size(y3));
...
```

We have two more layers to go: the pooling and convolution layers. The following listing shows the back-propagation that passes through the pooling layer-ReLU-convolution layer. The explanation of this part is beyond the scope of this book. Just refer to the code when you need it in the future.

```
...
e2 = zeros(size(y2));            % Pooling
W3 = ones(size(y2)) / (2*2);
for c = 1:20
  e2(:, :, c) = kron(e3(:, :, c), ones([2 2])) .* W3(:, :, c);
end

delta2 = (y2 > 0) .* e2;

delta1_x = zeros(size(W1));
for c = 1:20
  delta1_x(:, :, c) = conv2(x(:, :), rot90(delta2(:, :, c), 2),
  'valid');
end
...
```

The following listing shows the function Conv, which the function MnistConv calls. This function takes the input image and the convolution filter matrix and returns the feature maps. This code is in the Conv.m file.

```
function y = Conv(x, W)
%
%

[wrow, wcol, numFilters] = size(W);
[xrow, xcol, ~          ] = size(x);

yrow = xrow - wrow + 1;
ycol = xcol - wcol + 1;

y = zeros(yrow, ycol, numFilters);

for k = 1:numFilters
  filter = W(:, :, k);
  filter = rot90(squeeze(filter), 2);
  y(:, :, k) = conv2(x, filter, 'valid');
end

end
```

This code performs the convolution operation using conv2, a built-in two-dimensional convolution function of MATLAB. Further details of the function Conv are omitted, as it is beyond the scope of this book.

The function MnistConv also calls the function Pool, which is implemented in the following listing. This function takes the feature map and returns the image after the 2×2 mean pooling process. This function is in the Pool.m file.

```
function y = Pool(x)
%
% 2x2 mean pooling
%
[xrow, xcol, numFilters] = size(x);

y = zeros(xrow/2, xcol/2, numFilters);
for k = 1:numFilters
  filter = ones(2) / (2*2);    % for mean
  image  = conv2(x(:, :, k), filter, 'valid');

  y(:, :, k) = image(1:2:end, 1:2:end);
end

end
```

There is something interesting about this code; it calls the two-dimensional convolution function, conv2, just as the function Conv does. This is because the pooling process is a type of a convolution operation. The mean pooling of this example is implemented using the convolution operation with the following filter:

$$W = \begin{bmatrix} \dfrac{1}{4} & \dfrac{1}{4} \\ \dfrac{1}{4} & \dfrac{1}{4} \end{bmatrix}$$

The filter of the pooling layer is predefined, while that of the convolution layer is determined through training. The further details of the code are beyond the scope of this book.

The following listing shows the TestMnistConv.m file, which tests the function MnistConv.[10] This program calls the function MnistConv and trains the network three times. It provides the 2,000 test data points to the trained network and displays its accuracy. The test run of this example yielded an accuracy of 93% in 2 minutes and 30 seconds. Be advised that this program takes quite some time to run.

```
clear all

Images = loadMNISTImages('./MNIST/t10k-images.idx3-ubyte');
Images = reshape(Images, 28, 28, []);
Labels = loadMNISTLabels('./MNIST/t10k-labels.idx1-ubyte');
Labels(Labels == 0) = 10;    % 0 --> 10

rng(1);

% Learning
%
W1 = 1e-2*randn([9 9 20]);
W5 = (2*rand(100, 2000) - 1) * sqrt(6) / sqrt(360 + 2000);
Wo = (2*rand( 10,  100) - 1) * sqrt(6) / sqrt( 10 +  100);

X = Images(:, :, 1:8000);
D = Labels(1:8000);
```

---

[10]loadMNISTImages and loadMNISTLabels functions are from github.com/amaas/ stanford_dl_ex/tree/master/common.

```
for epoch = 1:3
  epoch
  [W1, W5, Wo] = MnistConv(W1, W5, Wo, X, D);
end

save('MnistConv.mat');

% Test
%
X = Images(:, :, 8001:10000);
D = Labels(8001:10000);

acc = 0;
N   = length(D);
for k = 1:N
  x = X(:, :, k);                    % Input,        28x28

  y1 = Conv(x, W1);                  % Convolution,  20x20x20
  y2 = ReLU(y1);                     %
  y3 = Pool(y2);                     % Pool,         10x10x20
  y4 = reshape(y3, [], 1);           %               2000
  v5 = W5*y4;                        % ReLU,         360
  y5 = ReLU(v5);                     %
  v  = Wo*y5;                        % Softmax,      10
  y  = Softmax(v);                   %

  [~, i] = max(y);
  if i == D(k)
    acc = acc + 1;
  end
end

acc = acc / N;
fprintf('Accuracy is %f\n', acc);
```

This program is also very similar to the previous ones. The explanations regarding the similar parts will be omitted. The following listing shown is a new entry. It compares the network's output and the correct output and counts the matching cases. It converts the 10×1 vector output back into a digit so that it can be compared to the given correct output.

```
...
[~, i] = max(y)
if i == D(k)
```

```
  acc = acc + 1;
end
...
```

Lastly, let's investigate how the image is processed while it passes through the convolution layer and pooling layer. The original dimension of the MNIST image is 28×28. Once the image is processed with the 9×9 convolution filter, it becomes a 20×20 feature map.[11] As we have 20 convolution filters, the layer produces 20 feature maps. Through the 2×2 mean pooling process, the pooling layer shrinks each feature map to a 10×10 map. The process is illustrated in Figure 6-19.

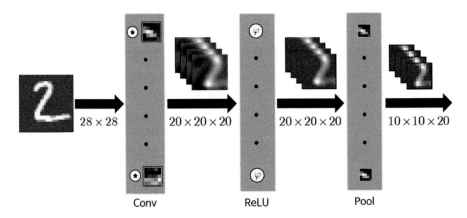

Figure 6-19. *How the image is processed while it passes through the convolution and pooling layers*

The final result after passing the convolution and pooling layers is as many smaller images as the number of the convolution filters; ConvNet converts the input image into the many small feature maps.

Now, we will see how the image actually evolves at each layer of ConvNet. By executing the TestMnistConv.m file, followed by the PlotFeatures.m file, the screen will display the five images. The following listing is in the PlotFeatures.m file.

---

[11]This size is valid only for this particular example. It varies depending on how the convolution filter is applied.

```
clear all

load('MnistConv.mat')

k  = 2;
x  = X(:, :, k);              % Input,        28x28
y1 = Conv(x, W1);             % Convolution,  20x20x20
y2 = ReLU(y1);                %
y3 = Pool(y2);                % Pool,         10x10x20
y4 = reshape(y3, [], 1);      %               2000
v5 = W5*y4;                   % ReLU,         360
y5 = ReLU(v5);                %
v  = Wo*y5;                   % Softmax,      10
y  = Softmax(v);              %

figure;
display_network(x(:));
title('Input Image')

convFilters = zeros(9*9, 20);
for i = 1:20
  filter           = W1(:, :, i);
  convFilters(:, i) = filter(:);
end
figure
display_network(convFilters);
title('Convolution Filters')

fList = zeros(20*20, 20);
for i = 1:20
  feature      = y1(:, :, i);
  fList(:, i) = feature(:);
end
figure
display_network(fList);
title('Features [Convolution]')

fList = zeros(20*20, 20);
for i = 1:20
  feature      = y2(:, :, i);
  fList(:, i) = feature(:);
end
figure
display_network(fList);
title('Features [Convolution + ReLU]')
```

143

```
fList = zeros(10*10, 20);
for i = 1:20
   feature     = y3(:, :, i);
   fList(:, i) = feature(:);
end
figure
display_network(fList);
title('Features [Convolution + ReLU + MeanPool]')
```

The code enters the second image (**k = 2**) of the test data into the neural network and displays the results of all the steps. The display of the matrix on the screen is performed by the function display_network, which is originally from the same resource where the loadMNISTImages and loadMNISTLabels of the TestMnistConv.m file are from.

The first image that the screen shows is the following 28×28 input image of a 2, as shown in Figure 6-20.

Figure 6-20. *The first image shown*

Figure 6-21 is the second image of the screen, which consists of the 20 trained convolution filters. Each filter is pixel image and shows the element values as grayscale shades. The greater the value is, the brighter the shade becomes. These filters are what ConvNet determined to be the best features that could be extracted from the MNIST image. What do you think? Do you see some unique features of the digits?

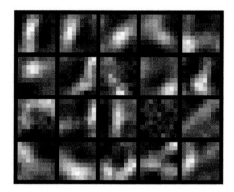

Figure 6-21. *Image showing 20 trained convolution filters*

Figure 6-22 is the third image from the screen, which provides the results (**y1**) of the image processing of the convolution layer. This feature map consists of 20 20×20 pixel images. The various alterations of the input image due to the convolution filter can be noticeable from this figure.

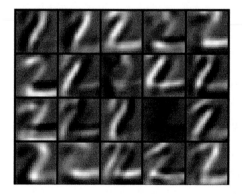

Figure 6-22. *The results (y1) of the image processing of the convolution layer*

The fourth image shown in Figure 6-23 is what the ReLU function processed on the feature map from the convolution layer. The dark pixels of the previous image are removed, and the current images have mostly white pixels on the letter. This is a reasonable result when we consider the definition of the ReLU function.

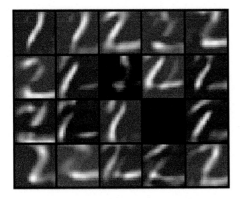

Figure 6-23. *Image showing what the ReLU function processed on the feature map from the convolution layer*

Now, look at the Figure 6-22 again. It is noticeable that the image on third row fourth column contains a few bright pixels. After the ReLU operation, this image becomes completely dark. Actually, this is not a good sign because it fails to capture any feature of the input image of the 2. It needs to be improved through more data and more training. However, the classification still functions, as the other parts of the feature map work properly.

Figure 6-24 shows the fifth result, which provides the images after the mean pooling process in which the ReLU layer produces. Each image inherits the shape of the previous image in a 10×10 pixel space, which is half the previous size. This shows how much the pooling layer can reduce the required resources.

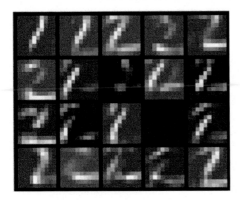

Figure 6-24. *The images after the mean pooling process*

Figure 6-24 is the final result of the feature extraction neural network. These images are transformed into a one-dimensional vector and stored in the classification neural network.

This completes the explanation of the example code. Although only one pair of convolution and pooling layers is employed in the example; usually many of them are used in most practical applications. The more the small images that contain main features of the network, the better the recognition performance.

Summary

This chapter covered the following concepts:

- In order to improve the image recognition performance of Machine Learning, the feature map, which accentuates the unique features, should be provided rather than the original image. Traditionally, the feature extractor had been manually designed. ConvNet contains a special type of neural network for the feature extractor, of which the weights are determined via the training process.

- ConvNet consists of a feature extractor and classification neural network. Its deep layer architecture had been a barrier that made the training process difficult. However, since Deep Learning was introduced as the solution to this problem, the use of ConvNet has been growing rapidly.

- The feature extractor of ConvNet consists of alternating stacks of the convolution layer and the pooling layer. As ConvNet deals with two-dimensional images, most of its operations are conducted in a two-dimensional conceptual plane.

- Using the convolution filters, the convolution layer generates images that accentuate the features of the input image. The number of output images from this layer is the same as the number of convolution filters that the network contains. The convolution filter is actually nothing but a two-dimensional matrix.

- The pooling layer reduces the image size. It binds neighboring pixels and replaces them with a representative value. The representative value is either the maximum or mean value of the pixels.

Index

© Phil Kim 2017
P. Kim, *MATLAB Deep Learning*, DOI 10.1007/978-1-4842-2845-6

Get the eBook for only $5!

Why limit yourself?

With most of our titles available in both PDF and ePUB format, you can access your content wherever and however you wish—on your PC, phone, tablet, or reader.

Since you've purchased this print book, we are happy to offer you the eBook for just $5.

To learn more, go to http://www.apress.com/companion or contact support@apress.com.

Apress®

Printed in the United States
By Bookmasters